Understanding Global Environmental Politics

Also by Matthew Paterson

ENERGY EXPORTERS AND CLIMATE CHANGE POLITICS (*with Peter Kassler*)
GLOBAL WARMING AND GLOBAL POLITICS

Understanding Global Environmental Politics

Domination, Accumulation, Resistance

Matthew Paterson
Senior Lecturer in International Relations
Keele University
Staffordshire

First published in hardcover 2000

First published in paperback 2001 by
PALGRAVE
Houndmills, Basingstoke, Hampshire RG21 6XS and
175 Fifth Avenue, New York, N. Y. 10010
Companies and representatives throughout the world

PALGRAVE is the new global academic imprint of
St. Martin's Press LLC Scholarly and Reference Division and
Palgrave Publishers Ltd (formerly Macmillan Press Ltd).

ISBN 0–333–65610–5 hardback (*outside North America*)
ISBN 0–312–23090–7 hardback (*in North America*)
ISBN 0–333–96855–7 paperback (*worldwide*)

This book is printed on paper suitable for recycling and
made from fully managed and sustained forest sources.

A catalogue record for this book is available
from the British Library.

The Library of Congress has cataloged the hardcover edition as follows:
Paterson, Matthew, 1967–
    Understanding global environmental politics : domination,
    accumulation, resistance / Matthew Paterson.
      p. cm.
    Includes bibliographical references and index.
    ISBN 0–312–23090–7 (cloth)
    1. Environmental Policy. I. Title.
    GE170 .P38 2000
    363.7—dc21

                         99–053110

10 9 8 7 6 5 4 3 2 1
10 09 08 07 06 05 04 03 02 01

Printed and bound in Great Britain by
Antony Rowe Ltd, Chippenham, Wiltshire

Contents

Preface

During part of my life as a student, we had a new Vice-Chancellor. Knowing that some of his academic work concerned the politics of environmental problems, I was optimistic concerning the possible openings for campaigns to improve the way the university dealt with its environmental impact. At that time (about 1991–92), the Green Society of the Students' Union was running a campaign concerning the promotion of car-use by the university. One part of this campaign concerned a plan by the university to spend £50 000, building a temporary (if I remember rightly to last only one year) car park to accommodate expanding demand for car-parking space, until funds and space could be found for a permanent one. We ('GreenSoc', as the logic of abbreviation by which all student political groups' names operate had it) produced a substantial document illustrating how, given the environmental consequences of car use, the university's limited resources would be better spent subsidising bus passes, working with the local council to improve cycle routes on to the campus, and so on. We timed this to coincide with the arrival of the new Vice-Chancellor, sent him a copy personally, on the assumption that what we needed to do was get his interest so that he would take up the matter with the relevant committee, and to provide him with alternative information to counter that produced by car-oriented bureaucrats in the planning office. However, in an ensuing issue of the university's official magazine, the new Vice-Chancellor printed a response to the campaign which effectively stated that the environmental problems associated with developments such as the car were determined by the logics of global capitalism and there was little point in the university spending its efforts or money in attempting to reduce the environmental impact of its decisions.

At that point, I realised I could never be a proper structural Marxist. At the same time, the reverse position, that all that is needed is to create sufficient political will for action so that a sustainable future can be forged, is equally unsatisfactory, even if less annoyingly complacent. What I hope to do in this book, is to highlight both how global structures of power systemically produce environmental change, but avoid the determinism and fatalism outlined above. An understanding of the structural constraints facing agents should not be understood to foreclose possibilities for action; rather it should precisely help to identify the possibilities for advancing social change. It is my hope that this book helps in such a project.

Acknowledgements

As usual, I owe many debts incurred in the production of this book. Deborah Mantle and Dave Scrivener both read the entire manuscript and provided invaluable feedback which substantially improved the flow of the text. Various people read individual chapters. Hidemi Suganami, Andrew Linklater, Debbie Lisle, and John Macmillan all read earlier versions of Chapter 4. Martin Parker read Chapter 6. John Macmillan, Simon Dalby and Cara Stewart also read Chapter 7. David Mutimer commented on parts of Chapter 3. A version of Chapter 4, combined with the general argument of the book, was given at seminars in the politics departments at Sussex University, Nottingham University and Staffordshire University, and the International Relations Department at Keele University. A version of Chapter 5 was presented at a seminar at Warwick University, at Carleton University, and at the British International Studies Association annual conference in 1998. A version of Chapter 6 was presented to MA students on the MA in Environmental Politics at Keele University. A version of Chapter 7 was presented at York Center for International and Security Studies in Toronto, and Trent University. I thank those present at these various seminars for helpful and stimulating discussions of the ideas developed here.

Parts of Chapter 3 appeared earlier as 'Green Politics', in Scott Burchill (ed.), *Theories of International Relations* (1996). I am grateful to Macmillan – now Palgrave – the publisher, for permission to reproduce this here. I am grateful to MIT Press for their permission to reproduce the figure on page 28.

Some of the research for Chapter 4 was carried out at Eastbourne Town Library. I am immensely grateful to the staff in the reference library there who were extremely generous with their time in finding many of the materials cited here. I am also grateful to Eastbourne Friends of the Earth and UK Friends of the Earth for letting me look at their files on their campaign on this question, and to those whom I interviewed: Peter Padgett, Eastbourne Borough Council Senior Engineer; Simon Counsell; Janet Grist and Mrs M. Pooley, Eastbourne Borough Council Environment Committee.

The Department of International Relations at Keele University has provided an extremely congenial and stimulating atmosphere in which to work, and I am grateful to all my colleagues for this. Dave Scrivener again merits particular notice for his timetabling genius, which makes it possible to make fairly heavy teaching commitments manageable. It is also important to recognise that my working in this way is ultimately parasitic on the labour of many women. The women at Kings Heath Grange Nursery who look after my daughter Freya during the day on low pay deserve mention – in particular Donna, Janet, Jo, Sarah, Rachel, Sonja, Annette, Laura, Lucy and Zoe. Linda and now Donna who clean up the mess after me, my partner, and Freya also make the life of a middle-class Western (male) academic less frenetic in a way which should not be left unacknowledged.

Finally, Underworld, Orbital (again), mix CDs from the Ministry of Sound and Cream, and Deep Dish have, among others, provided a soundscape within which I have immersed myself while writing this book. I have again been convinced that good dance-music is a great aid to writing.

1
Introduction: Understanding Global Environmental Politics

What does it mean to understand global environmental politics (GEP)? This question is deceptively simple. Behind it lies a set of thorny normative, theoretical and empirical problems. Increasingly, many people are challenging the ways of understanding GEP which have so far been dominant. These challenges from alternative perspectives disrupt what was previously a stable set of assumptions about what studying global environmental change within International Relations (IR) involved. Such stability however was not necessarily desirable, masking as it did the problems behind dominant perspectives.

To take perhaps an obvious question: what is it that we study when we study GEP? Clearly, an answer to this question is at least implicit in any understanding of such politics. It is customary, even ubiquitous, to begin books such as this with a discussion of the United Nations Conference on Environment and Development (UNCED), held in 1992 in Rio de Janeiro.[1] The discussion offers an evaluation of UNCED's outcome, and that of the process since then, and effectively sets up the debate in terms of how we explain the success or failure of the largest conference in human history.

Yet whether UNCED is interpreted as a failure (The Ecologist, 1993) or even as a step backward (Chatterjee and Finger, 1994), or with much greater optimism (Keohane, Haas and Levy, 1993), this construction of what the subject matter of GEP *is*, is already tendentious. It suggests that *the* question which students of GEP should ask is: what affects the possibility of states collaborating successfully to resolve particular transnational environmental problems? The understandings of GEP it offers are immediately a vision of the anarchic world of sovereign states, drawn into patterns of cooperation (and conflict) with each other over transnational problems. The fundamental logic of GEP is one of collective action. Specific research questions then flow from this which abound in the literature: the relative/absolute gains debate; the role of international institutions; the value of concentrations of power; the role of various groups (policy entrepreneurs, NGOs, scientists, and so on) in affecting the possibilities of cooperation. The immediate focus on UNCED has the effect of narrowing the terms of debate in ways which, I hope to show, are politically dangerous. As Doran (1993) quotes Guy Debord in opening his analysis of UNCED: 'The spectacle is the guardian of sleep.' To focus attention on the world's largest-ever diplomatic gathering is to focus on the spectacular, thus dulling the senses of observers to the political problems and dynamics such an event in name exists to discuss.

The last in this list of research questions, on the role of NGOs, could be seen to broaden the discussion away from a state-centric perspective. Certainly, in the work of some, it does (for instance, Princen and Finger, 1994; Wapner, 1996). But such groups are only usually considered in relation to the way they affect *interstate* collaboration (for example, Conca, 1995). The fundamental (yet largely unacknowledged, and certainly unexamined) commitments in this understanding of GEP are of an interstate understanding of global politics, of a liberal understanding of political economy, of the neutrality of science. So both social movements and their transformative potential, *and* transnational capital and its structural power, are reduced to 'NGOs' and their impacts on governments in particular contexts. Similarly, science, perhaps the ultimate rationalising discursive force of modernity, is reduced to the activities of scientists and research managers, 'epistemic communities', in relation to discrete problems of environmental change.

Not only is this way of framing what GEP is and how we should understand it very narrow in its construction, it also sets up the question in such a way so that only certain answers are permitted. If we ask how states collaborate to resolve environmental problems, then we are effectively precluded from answering that states are themselves (or alternatively, the states system is itself, through generating certain practices on the part of states) prime environmental destroyers. There is no space for such a view to be held. There are, however, many working within this tradition who do suggest that states are environmental degraders. But the contradiction between this view and the focus on interstate cooperation to resolve environmental degradation is not followed through.

Fundamentally, this framing is an attempt to *depoliticise* GEP. By constructing GEP in the language of collective action, it invokes a set of debates and literature (that is, rational choice and game theory) which represent political phenomena as technical ones (Paterson, 1995). It is depoliticising in the sense that it abstracts from questions of power, and it attempts (though it fails, as such attempts necessarily do) to exclude normative questions from the field of study.

From the point of view of the argument I develop in this book, perhaps the core failing of this way of framing GEP is to exclude the question of why global environmental change occurs in the first place. There is an implicit (and occasionally explicit) answer to this question. This answer is that global environmental change occurs as a result either of the 'tragedy of the commons [sic]', or of simply a set of disparate and discrete trends (which therefore require no explanation), as I show in detail in Chapter 2.

Yet an answer to this question is clearly logically prior to the question of how societies respond to such change. Only if it can be established that the central dynamic underlying global environmental change is in fact an interstate version of the tragedy of the commons, does it make sense to start asking how states attempt to mitigate this tragedy.

I would argue that there are three primary questions which should be taken as central to the study of GEP.[2] These questions concern:

- the production of environmental problems (why do they occur?)
- the differential effects of environmental problems by a variety of categories (class, nationality, race, gender)
- responses to these problems (what should we do?).

Any understanding of GEP contains implicit or explicit answers to these questions. Figure 1 shows, in extremely simplified form, the variety of answers which are discussed throughout this book, and which I take to be the main ones existing. It is clearly intended only as a very rough guide to the sets of positions which exist.

The first two columns in Fig. 1.1 represent the arguments which dominate debates about global environmental change in International Relations. The third column represents debates which date primarily to the 'first wave environmentalism' of the late 1960s and early 1970s. I will not discuss such a position in great detail in this book, partly because for political reasons it does not constitute much of a target – it is a bit of a 'straw person' – but mostly because in many cases in practice it offers a simple negation of realist and liberal institutionalist arguments. As such, the arguments which apply to those writers apply also to the eco-authoritarians.

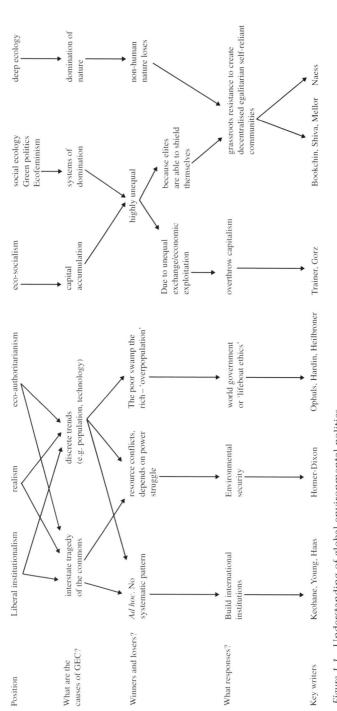

Figure 1.1 Understanding of global environmental politics

The bulk of the book is then intended to help develop the approaches on the right-hand side of Fig. 1.1. By contrast to the other positions, these are still largely uncharted territory for students of International Relations. In Chapter 3 in particular I introduce the ideas of these positions from the point of view of IR/global politics. However, I will also take these positions at a more generic level, and try to develop, through a small number of interpretive analyses, themes common to them all. These analyses are designed first of all to show that the answer to the first of my questions is not that global environmental change originates in interstate collective action problems, nor in a set of *ad hoc* trends, but in the internal dynamics of both systems of accumulation and exploitation, and systems of domination (both of humans by other humans, and of 'nature' by humans).

Collectively, therefore, the chapters which follow amount to an attempt to think through and develop existing critical approaches to GEP. There is now an emerging literature on GEP within International Relations which starts from such a critical perspective.[3] This literature, in differing ways, is critical of liberal and realist approaches to the study of environmental change within International Relations. The main intention of most of this literature is to disrupt the notion that international power structures are neutral with respect to environmental change. The liberal institutionalists who dominate the study of environmental change within IR assume such neutrality as they simply analyse the responses of states collectively to such change. Implicit (and occasionally explicit) is an assumption that the interstate system can *in principle* respond effectively to environmental change. Critical writers aim to challenge this assumption. Many of them (although not all) also wish to challenge the assumption that the states system is the only power structure on which it is relevant for students of IR to focus their attention.

However, the tendency in this literature is simply to argue that the international power structures are inconsistent with principles of sustainability, in the sense that they provide insuperable obstacles to achieving that goal. This could be, for example, because of the spatial mismatch between state sovereignty and the global scale of environmental change, or that the commitment to a deregulated globalising economy overrides attempts to regulate economies to pursue sustainability (for example, Elliott, 1998, p. 4).[4]

A stronger critical argument would be based on establishing that these power structures systemically *produce* environmental change in the first place, rather than simply preventing successful responses to that change. This would establish a stronger case for critical approaches than simply to argue that (for example) states cannot effectively respond to the environmental crisis because of the problem of the spatial mismatch. It also allows us to engage in interpretive analysis grounded in concrete processes and experiences, and suggests lines of enquiry which would enable us to develop more nuanced arguments about precisely what it is about capitalism (for example) that is unsustainable, and therefore what the necessary features of a sustainable political economy would be.

Julian Saurin (1994, 1996) is one writer who has tried to develop such an argument. Saurin suggests that rather than analyse how states 'have responded to the impact of environmental change – where the change is taken as given and relatively unproblematic – a thorough analysis of *causes* and of the *diffused* processes which engender environmental change' should be developed (1996, p. 79).

Saurin's argument here has two consequences. Firstly, it suggests that investigating such causes and diffused processes leads us to challenge the implicit assumption underlying the arguments of regime analysts, and in many ways also of writers such as Elliott, that environmental change is a 'consequence of accidents, errors or misunderstandings'. Rather, we should talk of the '*production* of environmental degradation' (Saurin, 1996, p. 81). This leads us to discuss the material practices which produce such change. 'Attention paid to globalised reiterated practices reveals incomparably more about the organisation and administration of degradation than does a focus on the ad hoc and tangential witnessed in inter-state environmental negotiations'.[5]

Secondly, his argument leads us away from discussing environmental 'issues'. As Saurin suggests in his critique of Steve Smith's provocative article about why the environment is 'on the periphery' of International Relations, this marginalisation stems at least in part from its conceptualisation as a set of discrete 'issues' – climate change, toxic waste, species extinction, and so on – which serve to marginalise the study within the discipline (Smith, 1993; Saurin, 1996, p. 78). But reducing 'global environmental change' to 'environmental issues' also serves to make each 'issue' appear discrete and by inference manageable, more amenable to technological fixes. This therefore abstracts from the systemic production of such change.

Later in the book, I develop this argument by examining social practices surrounding the car, and also meat-eating. Car production and use and meat-consumption clearly generate environmental change across a range of 'issues'. Simultaneously, both the car and meat-consumption have increasingly, albeit in some countries more than others, come under attack. There therefore exist practices of resistance to car culture, road building, or intensive meat-eating.

This book tries therefore to go beyond a critique of regime theory, and to provide an empirical analysis consistent with the basic principles of a Green position in IR. I take as my point of departure Saurin's core question, 'If degrading practices occur as a matter of routine, how do we account for this?' (1996, p. 90). I therefore ask the question How are the power structures of global politics implicated in the way that environmental change is generated? To answer this, I examine a set of social practices which systemically generate environmental change, and the way these practices are structured politically.

What I want to show in these later chapters is that both cars and meat are deeply embedded in the reproduction of global power structures. These daily consumptive practices and experiences simultaneously both systemically produce environmental degradation on global and local scales and also help to reproduce capitalist, statist and patriarchal identities and structures. I argue that such structures are deeply implicated in the production of environmental degradation. I organise this discussion around four themes: the relationship between the emergence of the car and the pursuit of capital accumulation; the promotion of the car by states; the role of the car in the reproduction of a variety of social inequalities; and the symbolic politics of identity whereby both cars and meat-eating are valorised as supremely modern (and rhetorically at least, therefore irresistible) commodities.

The following two chapters draw out in detail the different understandings of GEP illustrated in Fig. 1.1. Chapter 2 discusses realist and liberal institutionalist arguments. The intention is to establish that both realist and liberal approaches suffer from the shortcomings I have already alluded to. Specifically, they start from a position, derived from a particular (and rather narrow) conception of what International Relations *is*, where the relevant question is only what patterns of interstate responses to global environmental change can be observed, expected, and explained. At the inevitable risk of oversimplification, liberals focus on patterns of cooperation – the emergence of 'international regimes' to respond collaboratively to global environmental change, while realists generate the research agenda associated with the study of 'environmental security' – an approach which emphasises the potential of global environmental change to produce interstate conflict.

Both approaches therefore effectively assume that the first of my three questions (concerning the origins of global environmental change), and to a lesser extent the second (concerning the inequalities produced by global environmental change), are not relevant questions for students of IR to ask. Nevertheless, as Chapter 2 shows in more detail, they both have implicit assumptions about an answer to the first of these. In general, they assume that global environmental change occurs either because of an interstate 'tragedy of the commons', or as a result of a discrete set of what Keohane refers to as 'secular trends' (1993), such as increased consumption, population growth, technological change. However, such trends are not conceptualised explicitly, and thus the ways in which they might be systematically produced by global systems of domination and accumulation are missed. They are effectively conceptualised as accidents.

Chapter 2 also shows, however, that there were interesting developments within liberal IR theory during the 1990s, mostly associated with a shift from a conception of 'international regimes', firmly grounded in state-centrism, towards 'global governance', which is broader in scope. This trend is simply noted at this point; it is revisited in more detail in Chapter 7. Much of the literature associated with this shift still leaves my first question unaddressed, and while I suggest that the arguments advanced in favour of a notion of global governance can be thought of as more consistent with my argument concerning the causes of global environmental change, they should still be considered after an examination of such causes has been carried out.

Chapter 3 introduces ideas from the variety of positions on the right-hand side of Fig. 1.1 (as mentioned earlier, the 'ecoauthoritarians' will not be discussed). All positions discussed in Chapter 3 have a more explicit understanding of the origins of global environmental change. All effectively conceptualise such origins in structural terms. By this I mean that they think of global environmental change as the product of a systematically produced and integrated set of practices, rather than as a set of discrete trends. They will differ in their emphasis on particular structures, with, for example (again at the risk of oversimplification), ecofeminists emphasising patriarchy, deep ecologists emphasising the ideological structure of modern scientific knowledge predicated on the 'domination of nature', social ecologists emphasising political/social structures of domination, or ecosocialists emphasising the structural nature of capital accumulation.

Chapters 4–6 then attempt to illustrate the conceptual argument of Chapters 2 and 3 through three case-studies. It may seem odd, even indefensible, to base the book on cases drawn exclusively from the UK. Clearly, there are always problems in generalising from such analyses. In defence, I would answer primarily that quite apart from the pragmatic aspects of ease of organising the research, such a focus is consistent with the argument I make throughout the book, empirically and theoretically. I suggest that we should begin by focusing on the practices which produce global environmental change, which are necessarily 'local'. Normatively, since I argue, with many others, for a scaling-down of economic and political activity, a local focus is again legitimate. Finally, however, it is also the case that while the instances I discuss are drawn from the UK, the process of globalisation means that such practices and institutions (sea defences, cars, McDonalds) are 'local everywhere', to borrow from the discourse of global managers.

Through an analysis of the politics of sea defences, Chapter 4 shows how political decisions concerning global environmental change are deeply embedded in the broader reproduction of state, political-economic, scientific-technological and patriarchal power structures. The neutrality of political decision-making cannot be presumed, and thus liberal institutionalist positions which assume such neutrality are disrupted. Chapter 4 effectively starts from a focus generated by liberal institutionalist theoretical assumptions (albeit in perhaps an idiosyncratic fashion), and then proceeds to challenge the analyses produced by such theoretical lenses.

Chapters 5 and 6 then move to the question of the causation of global environmental change. Chapter 5 examines a set of social practices which generate such change, the practices surrounding the car. Likewise, Chapter 6 examines the ecological consequences both of the emergence of a fast-food culture, and of intensive meat-consumption. They show how those practices simultaneously produce global environmental change in a systematic fashion, and also help to reproduce the major power structures of world politics. Chapter 6 in particular also returns to questions of political agency, but in a vein more consistent with the argument of Chapter 5. It examines practices of ecological resistance to the systems of power of world politics, focusing on the McLibel case. I use this case to show how attempts to alleviate the social-ecological consequences of modern societies are necessarily bound up with attempts to radically alter the organisation of those societies.

In Chapter 7, I conclude by engaging with debates concerning political responses to global environmental change. It takes as its point of departure the emerging literature on global environmental governance (for instance, Lipschutz, 1997; Wapner, 1997), primarily because such literature does not restrict itself to questions of interstate collaboration. I suggest, however, that since the positions of Wapner and Lipschutz emerged conceptually out of the pluralism of the 1970s, they pay insufficient attention to how such emerging practices of governance are resistive to dominant forces in global politics. Chapter 7 therefore reorients these arguments in the light of the previous chapters, suggesting that political agency concerning global environmental change is necessarily about resisting globalising capitalism and building small-scale societies based on egalitarian social principles and a steady state economy.

2
Realism, Liberalism and the Origins of Global Environmental Change

Introduction

This chapter examines the research agenda of liberal and realist IR theorists in studying global environmental change. My first aim here is simply to outline the analyses they make of, on the one hand, international environmental regimes and, on the other, environmental security. I return to the question of political responses in particular in Chapter 7 where I will revisit these arguments and assess their value given my argument about the origins of global environmental change. In this chapter my concerns are more modest: to provide an account of the questions asked, assumptions made and themes developed by the majority of writers in IR who investigate global environmental change.

After such an exegesis, I attempt to show how mainstream approaches within IR to global environmental change exclude questions concerning the causes of global environmental change. This is not a huge claim, since most writers in these traditions would perhaps explicitly suggest that such questions are not part of their domain. But it remains nevertheless an important part of my task in the book as a whole. My aim is effectively to show that while they exclude this question from their field of enquiry they necessarily have an implicit assumption about what the causes of global environmental change are. In later chapters I argue that the implicit assumptions they hold concerning these causes are implausible. Here I merely wish to establish what their assumptions are. I suggest that they assume that the origins of global environmental change are in either (a) an interstate 'tragedy of the commons', and/or (b) a set of secular trends which are treated as exogenous to any conceptual or theoretical enquiry.

The realist and liberal global environmental change research agenda

Liberal institutionalism

The liberal institutionalist research agenda concerning global environmental change has focused primarily on accounting for the emergence of international environmental regimes. This focus has been the product of a shift in liberal thought in International Relations from the end of the 1970s onwards from the pluralist focus on a multiplicity of actors in world politics, the decreasing utility of physical force, and so on, back towards a state-centric analysis of global politics. This vision is thus much closer to the position of realism, especially given the more or less simultaneous structural turn taken by some realists, associated in particular with Waltz's *Theory of International Politics* (1979). As a consequence of this theoretical shift, the 'problem' for liberal institutionalists in terms of transnational problems is almost necessarily a question of explaining collective action. Since global politics is understood as *international relations*, as a realm of sovereign states interacting in an anarchic setting, any social or political problems which transcend state boundaries are necessarily understood as collective action problems, or alternatively, as problems concerning the provision of public goods. Such goods, characterised in technical language by 'jointness of supply' (so that, for example, no country can single-handedly provide a stable climate globally) and 'non-excludability of benefits' (no country can insulate itself from the impacts of climate change, or make sure that only it benefits from a stable climate), must be resolved through collaboration. International regimes have been the descriptive device on which explanations of such collaboration have been centred.

The literature on international regimes, however, did not emerge primarily in order to explain international environmental politics from a liberal institutionalist position. Mostly, it emerged to explain patterns of interstate interaction on the global economy (see, for instance, many of the chapters in Krasner, 1983). Occasionally it was used to explain security regimes (for example, Jervis, 1983). Krasner's definition of an international regime has become ubiquitous, but is perhaps worth repeating here. Krasner defines a regime as a 'set of explicit or implicit principles, norms, rules, and decision-making procedures around which actors' expectations converge in a given area of international relations' (1983a, p. 2). Thus the first point to note about regime analysis is that regimes are not the same as specific agreements; nor are they synonymous with particular organisations. Regimes are usually regarded as a subset of institutions – while the latter are broad in scope and content, regimes are narrower, confined to particular issues ('a given area of international relations') (Young, 1989, p. 13). Thus it is possible to have regimes where there are no written agreements through which they can be defined, yet 'principles, norms, rules, and decision-making procedures' can be identified which make up the regime. And as institutions, they are not to be treated as the same thing as organisations (Young, 1989, p. 32; 1994, p. 2).

Stokke suggests that the focus of regime analysis has been on four sets of questions (1997, pp. 32–5). The first of these concerns how regimes are maintained. Struck by the resilience of regimes even after the conditions which led to their establishment have waned (the classic example in the literature being the Bretton Woods regime), regimes scholars have investigated the reasons for such durability. The second concerns how regimes are formed, the various factors which might influence the possibility of establishing regimes (such factors as are, for example, emphasised by Young, or Hahn and Richards: see below). The third question concerns the consequences of particular regimes. Stokke is not entirely explicit about what he takes this to mean, but it appears to be about 'the ability of states to coordinate behavior in mutually beneficial ways' (1997, p. 33). This is perhaps only slightly different from his fourth question, whether regimes are effective. This final category of questions is for Stokke the most complex, as effectiveness proves rather difficult to define. It can be taken to mean the resolution of the problem for which the regime was established. Alternatively, it can be interpreted in terms of its effects on actors' behaviour, an interpretation which again has a number of ways of being put into practice. Each of these approaches has problems, primarily of a methodological nature (for example, how does one judge whether a state's behaviour has changed because of the operation of a particular regime?). It also raises perhaps deeper political questions concerning the way in which 'effectiveness' in the literature defers to natural scientists as definers of environmental quality, and to states as the only institutions capable of delivering environmental quality (Paterson, 1995, pp. 213–14).[1]

Stokke also suggests, following a distinction made originally by Keohane (1989, ch. 7), that each of these questions has been approached from broadly two perspectives – what Stokke calls 'individualist' and 'sociological' respectively, although Keohane referred to them as 'rationalist' and 'reflectivist'. The first of these suggests that regimes can best be analysed in terms of the strategic interaction of instrumentally calculating agents. It is thus amenable to analysis using game theory, the branch of rational choice theory devoted to situations involving strategic action (i.e. where outcomes for A are dependent on actions by B, C, etc.). In this perspective, regimes 'matter' because of the ways in which they alter incentives facing states, change the patterns of information available to them, thus changing their behaviour. The second position suggests, by contrast, that the (re)production of social meaning is a driving force behind behaviour, that 'people act toward objects, including other actors, on the basis of the meanings that the objects have for them' (Wendt, 1992, pp. 396–7). In this perspective, institutions, including regimes, play a deeper role than that conceived by rationalists in structuring state identities and thus negotiating positions.

An alternative typology of regime analysis is provided by Young, who suggests that concepts of 'power', 'interests' and 'ideas' are central to most analyses of regimes, although Young himself is critical of these approaches (1994, pp. 84–98).[2] In this typology, 'power' is understood in interstate terms, with much focus on hegemonic stability theory, on the capacity of dominant states to secure cooperation from others, and on the necessity (or otherwise) of the existence of a hegemonic state in order to generate or sustain regimes. The 'interests' model is basically the same as the rationalist model identified (and advocated) by Keohane. Those focusing on 'ideas', or developing a 'cognitive model', argue that the driving force behind environmental regime-formation is the generation of new knowledge which transforms states' perceptions of their interests. Haas's epistemic community approach (1989; 1990a; 1990b; 1992) in which transnational scientific communities drive forward cooperation on environmental problems, is the most prominent form of cognitive model.

At a more micro level of analysis, another set of literature focuses on particular 'factors' which influence regime success (for instance, Hahn and Richards, 1989; Young, 1994, pp. 98–115; Sand, 1990). Hahn and Richards, for example, suggest that the likelihood of regime formation and effectiveness increases with (a) greater scientific consensus, (b) increased public concern, (c) perceptions of fairness by negotiating partners, (d) increased short-term political payoffs, and (e) the existence of previous, related agreements. It decreases with (a) the increasing costs of action, and (b) the increasing number of participants (1989, pp. 433–7).

Using regime analysis, a number of authors have tried to explain how interstate collaboration to respond to specific global environmental problems has emerged and developed. Examples include Haas's work on the Mediterranean Action Plan and on ozone depletion (1989; 1990a; 1990b), chapters by Parson on ozone depletion, Levy on acid rain, Haas on the Baltic and North Seas, Mitchell on oil pollution, and Peterson on fisheries management in Haas, Keohane and Levy's *Institutions for the Earth* (1993), Vogler's (1995) analysis of oceans, Antarctica, outer space, ozone depletion, and climate change, applications by Young of his approach to climate change (1994, ch. 2), fisheries and seabed mining (1989, ch. 5), nuclear accidents (1989, ch. 6) and various aspects of environmental politics in the Arctic (1989, ch. 7; 1994, ch. 3; Young and Osherenko, 1993), Rowlands (1995) and myself (1996) to climate change.

Regime theory is usually couched in the value-neutral language of positivist social science. The intention is that norms only influence the choice of object of study (for example, Keohane, 1989, p. 21). However, regime analysts are usually in practice committed to the idea that regimes are (a) benign, and (b) can *in principle* provide adequate resolutions of global environmental change. I have offered two different critiques of regime theory elsewhere which I summarise here (1995; 1996; 1996a). One is empirical in nature, the other theoretical and political.

Empirically, the assumptions held by regime theorists are untenable. Drawing on Keohane's rationalist/reflectivist distinction (which is itself problematic) the rational choice version of regime theory is simply empirically implausible. In climate change negotiations, for example, states do not behave rationally in the sense understood by rational choice theorists. Leaving aside the problem of ascribing the characteristics of an individual to organisations like states, states have simply not had clearly articulated, consistently ordered preferences with regard to climate change, which they have generated autonomously, and which they have rationally pursued. Their practices can be more plausibly interpreted as searching collectively, that is, intersubjectively, for new norms to help generate actions to respond to climate change (for more details of this argument and evidence, see Paterson, 1996, ch. 6; 1996a). To be sure, they have had various other state goals which have infringed on the development of such norms, such as promoting growth, economic deregulation, etc., but these goals have themselves been disrupted by (and are perhaps undergoing some sort of transformation because of) global warming, and it is by no means clear what state interests are with respect to global warming. This observation can be generalised to other facets of global environmental change.

This argument is akin to what Keohane terms 'reflectivist' positions on international institutions (1989, ch. 7). But this position is itself also unstable. Partly this instability is because Keohane lumped together a number of writers and perspectives into one, conflating for convenience what in practice are widely diverging perspectives. But even if a coherent position, which has perhaps come to be known more commonly as 'constructivism' (e.g. Adler, 1997), and associated most prominently with the work of Wendt (1987; 1992) can be identified, it still remains problematic. For me, this is primarily because such work still identifies the international system solely as an interstate system, and ignores other aspects of global politics, such as capitalism or patriarchy (Paterson, 1996, p. 180; also Samhat, 1997, p. 359), but also because the focus on norms and the reproduction of social meaning tends to undermine the positivist epistemological basis of IR theory to which Wendt, for example, remains committed (Paterson, 1996a; Kratochwil and Ruggie, 1986).

Theoretically and politically, regime analysis still suffers principally from being bound to three problematic characteristics (Paterson, 1995). It is committed to an overly restrictive notion of what constitutes international relations – the interactions of sovereign states in an anarchic environment. This limits the relevant questions which can be asked, primarily by focusing always on consequences for the international system, rather than on a diverse range of questions in which structures of global power are implicated, such as environmental ones. Related to this, it tends to have an (implicitly) liberal notion of political economy – that states and capital/markets are fundamentally separate spheres of social life, which interact contingently rather than in a systemic fashion. Thirdly, it is committed to a positivist notion of the purposes of social science, which again narrows the range of questions which can be asked, in ways effectively criticised by Cox in his account of 'problem-solving' theory (Cox, 1986; Paterson, 1995).

To summarise, the liberal institutionalist research agenda of global environmental change has to date been concerned only with identifying the conditions under which states in an anarchic international system can cooperate over global environmental change. Regimes are the descriptive device which has come to be used to characterise such cooperative efforts.

More recently, a literature developing a broader notion of 'global governance' has started to emerge from the liberal IR tradition. This notion is clearly broader than that of international regimes. Specifically, a notion of global governance is at least implicitly less state-centric than that of international regimes. It is used to invoke the possibility of broader shifts in global politics away from a world which can usefully be characterised as one of interstate anarchy, towards a situation where there are a greater multiplicity of actors, many of whom operate transnationally. This literature explicitly echoes the 'world society' pluralist perspectives of the 1970s (Burton, 1972; Banks, 1984; Keohane and Nye, 1977). In part it has emerged out of the regimes literature, in that it is still concerned with the question of how order is produced and maintained in a world without government, but increasingly the possibility is being taken seriously that sovereign states are not the only entities capable of fulfilling governance functions (Rosenau, 1990; Rosenau and Czempiel, 1992; Young, 1997; Wapner, 1996; 1997; Lipschutz, 1997; Newell, forthcoming 1999).

Some scholars have been developing such notions in relation to questions of global environmental change (Young, 1997). Paul Wapner's work on environmental NGOs is a good example of such a theme. Drawing on his own research on organisations like Greenpeace, Friends of the Earth and the WorldWide Fund for Nature, Wapner (1996) develops this idea in Young's (1997) edited volume *Global Governance: Drawing Insights from the Environmental Experience* (Wapner, 1997). Wapner develops the notion of a 'global civil society' (also Wapner, 1996; Lipschutz and Mayer, 1996; Lipschutz, 1997), which suggests that the functions of a civil society in mediating between state and citizen are evolving globally, through the practices in particular of such NGOs and social movements. The increasing interstate organisation of political life globally provides some sort of analogy to the state domestically, while the integrated world market means that the domestic–international distinction breaks down in that sphere (Wapner, 1997, pp. 72–7). These two developments provide the basis for the possibility of a global civil society emerging, but it is the practices of NGOs which constitute such a society. Wapner suggests that increasingly, the development of such networks are starting to fulfil some regulative/governance functions, either unintentionally (through the ways that organisations necessarily help to diffuse norms which regulate social life, see p. 79) or deliberately, by on the one hand affecting patterns of interstate governance, or by themselves setting up governance systems. He gives the CERES principles developed by Friends of the Earth and others as an example. These principles are a code of conduct through which the practices of multinationals can be monitored and audited, with the intended regulative effect occurring both through consumer and shareholder pressure (ibid., also Wapner, 1996, pp. 129–31).[3]

Questions remain within this literature concerning the extent of such governance functions being fulfilled by non-state actors. Young in particular is sceptical (in his concluding chapter in 1997b). I will return to this debate in Chapter 7. For the present, I will move on to discuss a realist agenda concerning global environmental change.

Realism and environmental security

From the same basic ontology of interstate anarchy, realists in IR generate a rather different research agenda from that of liberals. From a small number of subtle, but important and fundamentally different assumptions, realists generate a research agenda which focuses on the potential for global environmental change to produce interstate conflict. Baldwin (1993, pp. 4–11) suggests that there are six focal points which divide contemporary realist and liberal IR theory. The first of these is the nature and consequences of anarchy, with realists suggesting that anarchy requires states to be concerned primarily with their survival, while liberals argue that the threats to states' survival are usually not acute enough to allow this to dominate policy-making. The second concerns the prospects for international cooperation, with realists suggesting it is 'harder to achieve, more difficult to maintain, and more dependent on state power' than liberals (Baldwin, 1993, p. 5, quoting Grieco, 1993). Thirdly, realists and liberals diverge over what motivates state concerns – the 'relative–absolute gains' debate. Realists tend to suggest that states are concerned primarily with the gains they make from (for example) cooperative ventures relative to those of other states, while liberals tend to assume that states are concerned only or primarily with absolute gains made by themselves. Fourthly, realists tend to focus on national security, implying this is the most important state goal, while liberals tend to focus on questions of political economy. Fifthly, realists tend to suggest that it is not state intentions that matter in determining outcomes, but rather the distribution of power resources or capabilities between states. Liberals, by contrast, place more emphasis on the intentions of state decision-makers. Finally, the two schools differ over the degree to which international institutions and regimes affect outcomes in international politics. Realists suggest that they are epiphenomenal, while liberals argue they have become important in affecting patterns of international cooperation and conflict.

My own view is that it is the relative–absolute gains debate which is crucial in accounting for the different theoretical positions of neorealists and liberal institutionalists. As a consequence of the nature of international anarchy, realists argue that states must always be concerned primarily with their own security, and thus of the ways in which other states, even current allies, may in future be able to threaten that security. As a result, the focus on relative gains makes achieving cooperation much more difficult, as is well established within the game-theoretic literature (for instance, Snidal, 1991; Axelrod, 1984). So the possibilities of cooperation, the priority given to security in state goals, the importance of institutions and regimes, and to a lesser extent the relative importance of intentions or capabilities, all depend on this assumption concerning the motivation of states derived from the assumption concerning the implications of anarchy. A preoccupation with relative gains therefore makes it more difficult to get cooperation going than liberals tend to assume, and renders international institutions epiphenomenal to the production of outcomes in international negotiations.

While realists have focused much less attention on global environmental change than have liberal IR theorists, the perspective produces two sorts of research questions, flowing from its differences from liberal institutionalism outlined above. The first is simply a concern to suggest that liberals are overly optimistic concerning the possibilities of securing adequate levels of international cooperation on global environmental change, as in other areas. But there has been little focus on this argument, partly because liberals have bent over backwards to accommodate this criticism. This they have done primarily by making their analyses highly conditional (if *x*, *y*, *z* conditions hold, then we can expect cooperation to be easier to achieve than if they don't) (Keohane, 1993).

The second focus from a realist perspective can be seen in what has come to be known as the 'environmental security debate'. The immediate origins of this concept seem to lie more in attempts by environmentalists to move environmental problems up the political agenda. Yet such attempts have inevitably drawn them into contact with preexisting notions of security, which derive in great part from realist and geopolitical traditions ingrained in the practices of policy-makers. Much of the tale of environmental security discourse has concerned the tensions produced by this engagement. Some traditional policy-makers and military elites have tried to use 'environmental security' to bolster their power base in a post-Cold War world, while others have suggested that the link dangerously weakens the traditional focus of *national* security. On the other hand, some environmentalists reject the link as politically dangerous as it would lead to a militarisation of environmental policy, while many others, aware of this possibility of cooption and militarisation, have nevertheless attempted to extend the traditional conception of security to embrace questions of global environmental change.[4]

It is in its realist variant, however (although the term realist is rarely used), that the discussion of environmental security has perhaps had its strongest political impact. This is precisely because the term resonates with the interests of many in state elites, particularly within the military (concerned with their role and their resources after the collapse of their traditional enemy), and with dominant discourses concerning the priorities of state decision-makers.

In a realist mode, environmental security is simply an additional component to preexisting notions of security. The referent of security (what is to be secured) remains the same – the nation-state – while only the causes of insecurity have changed (from military enemies to environmental degradation). Some of these 'new' threats are old ones dressed up as environmental conflicts – the struggle between states for access to strategic resources. But in general the novelty is found in the potential for new conflicts over renewable resources (water, croplands, forests and fisheries are often identified), whereas the old conflicts were over non-renewable resources (oil, for example) (e.g. Porter, 1998, p. 217). By and large, however, the means of achieving security have in this discourse also remained largely the same, despite the shift in focus concerning the origins of insecurity. The institutions of the military are conceptualised as useful in responding to environmental change (for instance, Butts, 1994; Oswald, 1993). This is either because such change precipitates political conflicts of a conventional type, meaning that military force is required, or more commonly that the military has the technology, skills, and so on, to help ameliorate environmental problems (by helping with monitoring, for example).

There are two senses in which this discourse suggests that environmental change threatens security. The first is the claim that environmental change can lead to interstate war. Water is perhaps the most commonly cited resource over which wars could be fought. The predominant image in this literature is one where conflicts arise because two or more countries share a resource, for example a river basin, and one country is able to prevent others from getting access to the water on which they depend, and interstate conflict ensues. River basins such as the Nile, the Jordan, and the Tigris-Euphrates are all often cited as potential sites of such conflict, although the likelihood of conflict, and in the case of the Jordan the role of water conflicts in producing wars which have already happened, is heavily contested (see, for example, Bulloch and Darwish, 1996; Kliot, 1994; Hillel, 1995; Thomas and Howlett, 1993; Shaheen, 1997).

The second sense, emphasised in particular by the work of Homer-Dixon and colleagues (e.g. 1998; 1994) is that while interstate war may be an unlikely result of environmental changes, internal instability of states because of such changes is highly plausible. The starkest and highly popularised version of this argument is in Kaplan's rather apocalyptic vision (1994). Kaplan portrays a vision of a future based on an account of West Africa's present. He suggests that the root of the threat is '*nature unchecked*' (1994, p. 190, his italics). He continues:

> It is time to understand 'the environment' for what it is: the national security issue of the early twenty-first century. The political and strategic impact of surging populations, spreading disease, deforestation and soil erosion, water depletion, air pollution and, possibly, rising sea levels in critical, overcrowded regions such as the Nile Delta and Bangladesh – developments that will prompt mass migrations and, in turn, incite group conflicts – will be the core foreign policy challenge. ...
>
> (1994, p. 190)

While he does go on to suggest that wars could result from environmental degradation, the primary images of state insecurity he invokes are of internal decay and collapse, and of the decline in the relevance of borders, what he calls the 'lies of mapmakers'. Environmental degradation combined with (and also caused by: see more below on this) population growth produces rapid unplanned urbanisation, spreading disease and the breakdown of social order as states are unable to contain such developments.

Kaplan cites the work of Homer-Dixon (1991) as evidence for his thesis. Homer-Dixon's claims are rather more circumspect. Homer-Dixon, contrary to Kaplan's usage of his work, does not suggest that environmental degradation *will* cause or has caused wars (1998, p. 207) although he suggests a myriad of mechanisms by which such conflict *might* occur (e.g. 1991). In fact, he states that 'there is virtually no evidence that environmental scarcity causes major interstate war' (1998, p. 207). But in contrast, he suggests that there is much evidence that environmental degradation has caused major social conflicts within societies. He cites examples such as the Chiapas uprising in Mexico, conflicts in 'the Himalayas, the Sahel, Central America, Brazil, Rajasthan, and Indonesia' (1998, p. 206). Such conflicts are caused either through resource capture, where powerful social groups provoke conflict by 'using their power to shift in their favor the regime governing resource access' (1998, p. 205). Alternatively, they are caused by ecological marginalisation, where the poor are driven into ecologically marginal areas, which produces both greater ecological destruction and social conflict. Homer-Dixon doesn't go anywhere near as far as Kaplan in projecting this forward as an image of total social collapse, however, and in some places is relatively optimistic concerning the possibilities of managing and preventing such conflicts (e.g. 1998, pp. 210–11).

For this version of an environmental security agenda, then, the referent is clearly the nation-state. Environmental problems which do not adversely affect national security are therefore not of concern – as Elliott puts it, 'The problem is not environmental degradation *per se*' (1998, p. 220). In pieces like Kaplan's, the realist subtext lies not so much in a Waltzian certainty of the durability of the states system, but rather the reverse. Indeed, while Waltz lays a wager in *Theory of International Politics* (1979, p. 95) that the states in the world are almost certain to be around in 10 years' time, whereas the major multinational corporations are not, Kaplan is confident that they could well not be around in their current form, even in the case of states like the US.[5] But Kaplan's is still a realist concern. It has more in common with a (conservative) moralist version of realism which argues normatively that states are desirable political forms, providing the basis for individual security, prosperity, freedom, and so on. Threats to the stability of states should therefore be avoided.

But at the same time, the realist origins of environmental security reveal clearly its problems. As Dalby points out (1996), Kaplan's narrative of environmental degradation producing social collapse, producing transnational flows (drugs, migration, and so on) which are destabilising globally, masks the ways in which other transnational flows (such as trade, finance) themselves help to produce both the environmental degradation and social collapse which so worries Kaplan in the first place. The complicity of the 'West' in producing destabilisation in the 'South' is evaded.

Starting from basic assumptions similar to those of liberal institutionalists, but with small yet fundamental differences, realists generate a significantly different research agenda from that of liberals. They could develop an agenda (although there has been little effort yet to do so) which argued that liberal institutionalists writing about global environmental change were overly optimistic concerning the consequences of the plethora of international environmental regimes which have emerged since the 1970s. On the other hand they could generate, and in some ways have already done so, one version of the 'environmental security' debate. I now turn to examine how these two research agendas make assumptions (usually implicit) concerning the origins of global environmental change.

The causes of global environmental change in realist and liberal IR theory

The two clearest examples illustrating the assumptions of realist and liberal IR theory concerning the origins of global environmental change can perhaps be found in Vogler (1992) and in Litfin (1993). In a single paragraph alluding to such questions (he, like others working in the regime theory framework, states that the fundamental purpose is to analyse 'political responses' to global commons problems [Vogler, 1992, p. 118]), Vogler suggests the following. Firstly he states that political and economic problems to do with the commons 'arise from their rapid exploitation (made possible through the agency of technological change) and overuse' (ibid., p. 121). He then goes on in the same paragraph to discuss Hardin's 'tragedy of the commons' thesis explicitly. The consequence of this combination of *ad hoc* (technological) change and a collective action explanation of 'the problem' leads Vogler to formulate the problem as one of 'cooperative management' (ibid., p. 118). Litfin (1993, p. 105) suggests due to that the 'post-WWII increase in population and technological pressures, resources formerly perceived as vast or even unlimited are now seen as scarce and endangered. Consequently, the principles of open access and free use that formerly governed the global commons … have proven themselves inadequate to the point that "tragedy of the commons" has become virtually a household term.' This combination of a set of discrete trends and a background condition – the 'tragedy of the commons', underlies many discussions of environmental change in IR.

Tragedy of the commons

For many writers on global environmental change, the 'tragedy of the commons' metaphor is a useful model explaining the underlying permissive cause of such change. Most writers eschew the apocalyptic language of 'tragedy', but still invoke the notion of the *commons* as a metaphor for many facets of global environmental change. Young, for example (1989; 1994) suggests such a line of reasoning, citing the many analyses which have refuted Hardin's claim on empirical and theoretical grounds (for instance, Berkes, 1989; Ostrom, 1990; McCay and Acheson, 1987). The characterisation of global environmental change as a problem of global commons is ubiquitous (for example, Vogler, 1995; Rowlands, 1995, pp. 3–4; Keohane, Haas and Levy, 1993, p. 10; Soroos, 1997; Buck, 1998).

In this construction, the notion of the commons is used as a way of describing the nature of global environmental problems. But usually, the language of 'tragedy' is lost. They remain simply as problems of collective action, concerning the provision of public goods. Occasionally, Hardin's language is deliberately invoked as a characterisation of global environmental change (for instance, d'Anieri, 1995, p. 153; Brenton, 1994, p. 4; Hempel, 1996, p. 84). But even where writers reject Hardin's deterministic assumptions concerning the consequences of leaving resources unowned, the premise of an open-access resource with at least the *potential* to be overused remains; otherwise there is nothing either to be explained or to worry about. In other words, something of Hardin's image remains.

But in a stronger, yet usually less clearly articulated sense, the tragedy of the commons metaphor is taken as an explanation of the causes of global environmental change. In this sense, it is the absence of a global political authority, or, in the terms of a different yet related debate, the spatial mismatch between sovereign states and the global nature of environmental change (Conca, 1994), which acts as a permissive cause of global environmental change. In Hardin's original narrative, it is not so much that the absence of any property rights in the commons acts as an obstacle to cooperation, as focused on by Young and others, but more that such an absence is what creates the incentives for overgrazing in the first place, and therefore the need for cooperation. This is the meaning of Hardin's phrase 'freedom in a commons brings ruin to all' (1968, p. 1244). The herders overgraze *because* of the absence of property rights.

Those using Hardin's metaphor in relation to global environmental change tend to focus on the way in which the property relations in an open access resource (which Hardin erroneously called a commons; see, for example, The Ecologist, 1993, pp. 12–13) create obstacles to cooperation. Yet the causation of environmental change is an equally important part of Hardin's argument. This is a largely unacknowledged part of the logic of realist and liberal arguments.

Hardin's metaphor, and the notion of the commons as a cause of environmental change more generally, does however creep into the work of some authors. For Sandler (1997, pp. 11–12), environmental degradation occurs as a result of problems in allocating property rights (this is the standard formulation among environmental economists).

Such a market failure occurs when property rights are 'undefined or owned in common with unrestricted access' (1997, p. 11). It is the open access which is the problem here. Sandler's argument follows a now common modification of Hardin, rejecting his notion that all common property leads to his tragedy (Sandler, like Young, cites Ostrom, 1990). But he does argue that if a resource is to be owned in common, then there must be restricted access and regulated use. Similarly, Brenton (1994, p. 4) and Hempel (1996, pp. 84–5) both also mention the mixed empirical evidence for Hardin's thesis. Miller (1995, pp. 54–5) cites Hardin's metaphor both in terms of incentives to overuse resources and in terms of constraints on cooperation; like the others already cited, however, she departs from Hardin in suggesting that collaboration is possible under such conditions.

Nevertheless, the primary image of the commons in writing in IR on global environmental change is as a constraint to cooperation, and not as a cause of environmental change. I would suggest that this is at least in part because since Hardin's original argument ended up advocating authoritarian centralised solutions – 'mutual coercion, mutually agreed upon' (Hardin, 1968, p. 1247) – with world government as the corollary in global politics. Some writers do discuss Hardin's work in this context, discussing the tragedy as a model of the causes of environmental change (e.g. Wapner, 1996, pp. 28–9). Realists have always been sceptical concerning the possibility of achieving world government, and liberals, at least in the 'international regimes/global governance' literature, have increasingly moved towards an explicit argument concerning the possibility of 'governance without government'. Consequently, to emphasise the original logic of Hardin's argument would run the risk of directly contradicting their own normative arguments. If you aren't going to emphasise the way in which a system defined by the absence of property/sovereignty rights at the global level produces environmental change, then you have less of a problem in arguing that such a system is compatible with resolving such collective action problems. The debate can turn on the game-theoretic arguments concerning the possibility of 'cooperation under anarchy' (Oye, 1986; Axelrod, 1984; Taylor, 1987), already rehearsed above in relation to general debates between realists and liberal institutionalists.

Yet perhaps this is a deep contradiction within liberal writings in particular. Realists have fewer problems with this conclusion; they can simply argue that perhaps the logic of anarchy in international politics means that no successful responses to global environmental change will emerge. Whether this is pessimistic or not, they can refer to this as one of the 'realities' of international relations. But liberals are rather more optimistic in their assessment of the possibilities of such successful responses. At the same time, they have, at least since the retreat from the strong pluralist positions of the 1970s by Keohane, towards a position where interstate anarchy is, as for realists, taken as the defining feature of international politics (I take Keohane both to be an exemplar of such a shift, and in some sense a power-broker in it). So while, to the extent they relate the international system as they see it to causes of global environmental change, they must do this in terms more or less of a 'tragedy' metaphor. Perhaps this contradiction explains why such writers tend to focus on particular discrete trends as the causes of global environmental change, to the extent that they discuss them at all.

Discrete trends

Keohane, Haas and Levy give a set of 'factors' which produce environmental degradation. 'Many environmental threats are caused by such factors as population pressures, unequal resource demands, and reliance on fossil fuel and chemical products which degrade the environment' (1993, p. 7). In their conclusion to the same book, they again give a similar list. 'Each set of issues has been considered separately, independently of possible underlying causes such as population growth, patterns of consumer demand, and practices of modern industrial production' (Levy, Keohane and Haas, 1993, p. 423). Keohane, in a more general piece on liberal institutionalist theory, explicitly suggests that 'increased economic and ecological interdependence' are the result of 'secular trends' (1993, p. 285). While the context of this point is to provide a contrast with realist post-Cold War expectations concerning the demise of institutions, in particular the EU, it also shows how such things are regarded as having their origins outside the fundamental operation of the states system.

Lorraine Elliott gives a list couched in similar terms, albeit in less 'value-free' language. She writes, 'Extensive and excessive resources use, energy-inefficient lifestyles, industrialisation and the pursuit of economic growth are inextricably linked to environmental degradation' (1998, p. 1). In more generic language, Sandler writes that environmental degradation is the result of 'the actions of individuals to keep warm, to feed themselves, and to produce goods and services', which will be exacerbated as a result of population growth in developing countries (1997, p. 2).

Oran Young, despite his extensive writings on the subject, manages to stay remarkably clear of explicit discussions of the causes of environmental change. In his *International Governance* (1994), the closest he gets is in stating that 'human behaviour is a critical driving force' (p. 51) which is rather underspecific, to say the least. He then complements this by saying such behaviour has its origins in the notion of 'nature as resource', there for human instrumental use (ibid., citing Lynne White Jr, 1967). In *International Cooperation* (1989), he suggests that cooperation will become more elusive concerning environmental change 'as growing human populations, enhanced capabilities, and rising expectations generate more severe conflicts of interest as well as greater demands on the earth's natural systems' (p. 4, see also pp. 107–8). Some indications as to causes are perhaps implicit in his threefold typology of environmental problems into commons problems, shared resources problems, and transboundary externality problems (1994, pp. 20–4; 1997, pp. 7–8). But they are not developed in his work.

The most extensive treatment in this literature is in the first few chapters of Nazli Choucri's edited volume *Global Accord: Environmental Challenges and International Responses* (1993). For Choucri, the causes of environmental change are understood in profoundly individualist and social-psychological terms. 'The most fundamental unit ... is the individual human ... who ... responds to felt needs, wants, and desires, by making demands and acting upon natural and social environments in order to obtain the sustenance without which he/she cannot long survive' (Choucri, 1993a, pp. 9–10). This is rather like Sandler's account given above, where the origins are in the meeting of human needs and wants. But Choucri then builds on this to offer an account of why environmental change has accelerated in the past century. She suggests it is the 'intended or unintended consequences to nature resulting from human action taken in pursuit of narrowly defined human interests' (ibid., p. 2). The specificity regarding modern environmental problems is explained as a result of 'human knowledge and skills (technology) interacting with population trends and demands for resources (and derivatives thereof) [which] have generated environmental problems worldwide' (ibid., p. 1). It is these three – population trends, technology, resource access – which Choucri fixes on as causes of environmental change (ibid., p. 13).

In the context of an attempt to chart possible scenarios of environmental change through to 2025, Thomas Homer-Dixon systematises the relationship between these three into a diagram, reproduced as Fig. 2.1 (Homer-Dixon, 1993, p. 44). Homer-Dixon suggests that the factors at the top of the diagram (institutions, social relations, beliefs

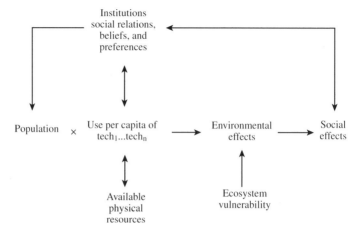

Figure 2.1   Main variables and causal relationships
Source: Choucri, Nazli (ed.) *Global Accord: Environmental Challenges and International Responses*, Cambridge MA: MIT Press (1993).

and preferences) are the most important in determining likely trajectories of social and environmental change. But he refers to these as 'ideational factors' (p. 45), He continues: 'This *social and psychological context* is immensely complex', and then lists a series of such factors: land distribution; wealth distribution; economic, legal and political incentives for consumption (including property rights and markets); family and community structures; patterns of trade; coercive power; metaphysical beliefs about nature (ibid., p. 45). In what sense can most of these be regarded sensibly as ideational factors? Surely, it makes more sense to regard them as structural factors related to the way societies are organised? I will revisit this point later on.

   Homer-Dixon then details particular trends which will be crucial over the next few decades. He mixes in these particular environmental trends (climate change, ozone depletion, loss of fisheries and agricultural land, deforestation, reduction in water supply and quality, decline in biodiversity) with trends in human society (population growth, energy consumption). These are intended to be particular examples of more generic phenomena described in the figure, but whether they are intended to have any explanatory capability is unclear.

   In the realist discussion of environmental security, the overriding dynamic is often taken to be demographic. This can be seen in Kaplan (1994) and Homer-Dixon (1998), for example. For the former, population growth and migration are almost taken to be examples of environmental change in themselves. For Homer-Dixon, population is taken formally to be one of three sets of dynamics (alongside the physical vulnerability of the resource in question, and the technological and consumptive practices of the relevant populations), but it remains a highly emphasised theme. The focus on population clearly feeds into the political problems of the realist environmental security agenda. There is much in common with the old geopolitical imagery of the West being 'swamped' by the poor of the South and, as is often argued by those critical of too heavy a focus on population growth, it tends to evade the responsibility of the West, since the 'problem' appears to be produced elsewhere, by someone else.

Towards a structural account of global environmental change

One of the effects of operating with these two assumptions concerning the origins of global environmental change is to take existing systems of power for granted. Firstly, they do this by reducing all relevant forms of power to the operations of the interstate system.

But this effect operates differently for each of the two forms of assumption. Concerning the tragedy of the commons, the assumption is that global environmental change is in part caused by the interstate system – because it promotes overexploitation of open access resources – but the interstate system is not put into question as a consequence (except by ecoauthoritarians, who take the same assumption and follow its logic through). The question becomes one of how international regimes or patterns of global governance mitigate this competitive, overexploitative dynamic. The tragedy of the commons is thus treated either as a conditional consequence of the state system (one which could or could not produce global environmental change, depending on various factors), or as a necessary consequence of the system, but where concerns about the environmental consequences of the state system are not allowed to override the ontological or normative priority of that system.

For the second assumption, by contrast, the trends are presented as *ad hoc*, not considered as consequences of power. This treatment is at least in part because power is reduced to its interstate forms, although many of the trends considered could be seen more as consequences of other power systems, for example capitalism.

On occasion, this can be associated with the attempt to depoliticise the analysis, as I suggested in Chapter 1. Oran Young (1997a) provides much material to support such a claim. Young suggests, for example, that one of the features of studies of international regimes and/or global governance is that they attempt to be of direct 'value to policy-makers' (ibid., p. 1). Immediately, the politics of environmental change is defined as one of management, with 'policymakers' as neutral or (potentially) benign managers. As Saurin suggests, this constitutes a reduction of agency to policy, which is deeply politically problematic (1996, p. 94). Later, Young offers a critique of Keohane and Nye's model of regime change (1977). He suggests there that 'It overempha-sized material conditions in contrast to shifting ideas and configura-tions of interests; reflected a preoccupation with power, which is common among students of politics but which is misplaced in the study of international or transnational regimes' (Young, 1997a, p. 18). A clearer attempt to depoliticise the subject would perhaps be difficult to find. The later chapters of this book will hopefully help to dispel the notion that global environmental politics can ever be free from ques-tions of power; here perhaps it will simply be noted that Young's posi-tion is in great danger of obscuring the deep political questions concerning global environmental change through attempts to define it in neutral terms.

On other occasions, writers in mainstream IR traditions do go some way towards a structural understanding. Take the introductory chapter of *International Organizations and Environmental Policy*, for example (Bartlett, Kurian and Malik, 1995). Written by the editors (Kurian, Bartlett and Malik, 1995), this chapter starts to move towards the argu-ment I will be detailing later, but seems disciplined in a curious man-ner by (North American) IR. This results in some heavily contradictory arguments. They outline their understanding of ecological rationality, which 'is evident in decisions and actions that result in maintaining a sustainable relationship between humans and the environment' (ibid., p. 4). They then argue not that ecological rationality requires a steady-state society, but that 'ecological rationality necessarily needs to take priority to attain some form of steady-state society' (since it is the condition of possibility of other sorts of rationality) (ibid.); the need for a steady-state society comes from somewhere else, unex-plained except for the first part of the sentence which places the above quote in 'the context of ecological scarcity', citing Ophuls and Boyan (1992).

The authors continue with arguments which are in many ways con-sistent with the one developed in Chapter 3. There is a focus on eco-nomic growth, and the way international organisations and states are enmeshed in a 'system that is linked by the imperative of economic growth' (ibid.). There is an implicit assumption that growth is environmentally problematic, although the closest the authors get to trying to argue this is in the use of a quote from Michael Redclift (1987, p. 56, cited in Bartlett, Kurian and Malik, pp. 4–5): 'resource depletion and unsustainable development are a direct consequence of growth itself'. There is also a focus on the state and state system as environmentally problematic (ibid., pp. 5–6); on science (10, 12), and on the contribution of feminist perspectives (15).

But what is important for the present purposes are two questions. First is the question of what lies underneath particular identified dynamics (growth, administrative rationality of the state, and so on). To continue with the discussion of economic growth, the authors offer ambiguous arguments concerning whether growth is a systemic imperative or a simple goal which countries aim at. They write that 'the pressures of economic systems are such that, even in the unlikely event of nations wishing to stay out of the scramble for growth, they cannot do so; the coercive reality of survival and the interdependence of all nations ensure the continuation of an environmentally destructive system' (ibid., p. 5, citing Walker, 1989; Redclift, 1987). But just prior to the passage quoted it is argued that both 'capitalist and socialist systems' have 'focused efforts on maximum exploitation of nature', as a *consequence* of 'notions of resource abundance'. In other words, the growth imperative is produced by ideologies of nature rather than systemic pressures.

A second question is, however, how the focus of the whole book ends up with an almost exclusive focus on international organisations, after the intimations of something more critical in the opening chapter. The contradiction between the opening focus on the problematic nature of major systems of power with respect to environmental problems, including statements such as '[governments] react to symptoms but seldom to causes which tend to be regarded with suspicion as possibly leading to "subversive" changes' (ibid., quoting King and Schneider, 1991, p. 4), and the later focus on international organisations and the policies they pursue, is striking. It seems to be produced by the intellectual and disciplinary orthodoxy of IR, especially in North America. But if the arguments outlined at the book's beginning are taken seriously, and especially if the structural, systemic understandings in them are emphasised, then much of the rest of the book is little more than irrelevant, since its basic assumptions have been undermined.

Hempel (1996) also provides an example of an argument which *almost* provides a structural account of environmental problems. Certainly his aim seems to be to develop such an account:

> Rather than attributing environmental destruction to the actions of a relatively small number of thoughtless and careless individuals, or to some passing phase of industrial recklessness that accompanies an otherwise benign evolutionary process of economic development, the destruction described here is attributed to driving forces that are pervasive, persistent, and deeply ingrained in our values, lifestyles, and institutions.
>
> (Hempel, 1996, p. 52)

Hempel reviews the debate concerning underlying causes of environmental degradation which is often expressed in the equation $I=PAT$ (ibid., pp. 58–60).[6] He suggests that such a numerical model is inevitably oversimplified, and proposes a qualitative model involving what he calls eight driving forces. These forces are grouped into four types, organised hierarchically. He begins with core values, which are anthropocentrism and contempocentrism. His second set are amplifiers, which are population and technology. The third set is consumptive behaviour, comprising poverty and affluence. Finally, there is political economy, involving on the one hand market failure, and on the other, the failure to have markets (ibid., pp. 60–86).

While such a focus is systemic to a degree, in an important sense Hempel's analysis misses the point of a systemic or structural argument. His driving forces are all ultimately presented as if somebody chose those core values, population levels, or whatever, and as if we could simply choose different versions of the eight forces which are consistent with ecological principles. The point which is missed is the systemic logic which produces these forces in the first place. From this perspective, it is not a historical accident that anthropocentric ethics emerged at the point where (at least in north-west Europe) modern scientific rationality was being developed, where patriarchy was being reconsolidated, where modern nation-states were emerging, and where capitalism was starting to take hold as the increasingly dominant mode of production. The same could be said for many other of Hempel's forces.

Hempel's failure to see the systemic logic underlying his forces produces particular contradictions. An important one concerns the connections between particular forces. In particular, the focus on market failure and failure to have markets comes from a neoclassical perspective where markets themselves are not problematic. He hints at a critical perspective when he states that 'some environmental critics' have concluded that 'the real economic driving force behind environmental destruction is neither market failure nor the failure to have markets, but rather the presence of markets that are working *too well*' (ibid., p. 83, emphasis in the original). But, in general, he follows a conventional line in arguing that we need to make markets work better to alleviate environmental problems (especially concerning applying markets to particular resources to 'price' them).

However, the points about markets are seen as unconnected to the points about poverty and affluence. From a structural perspective, the important point is not to talk about markets in the abstract, but historically produced capitalist markets, which have engendered poverty and affluence on a global scale, and the environmental degradation associated with them.

Similar allusions are made by Keohane, Haas and Levy (1993, p. 7) and by Elliott (1998, p. 1). The former suggest that 'while environmental degradation is ultimately the result of aggregated individual decisions and choices, individual choices are responses to incentives and other forms of guidance from governments and other national institutions via laws, taxes, and even normative pronouncements' (Keohane, Haas and Levy, 1993, p. 7). This concedes some ground to a structural perspective, but not a great deal. Couched in heavily rationalistic tones, it implies that political structures produce individual decisions in a stimulus–response manner – choices are 'responses to incentives'. If structure is conceived of as constitutive of identity, rather than merely as an external constraint, then its consequences are significantly different.

Conclusion

This chapter has shown how the two dominant perspectives in IR generate particular research agendas for studying global environmental change. Both start by asking how the international system responds to such change. Both assume that the system is best characterised as sovereign states interacting in a condition of anarchy, but have different assumptions concerning the consequences of the assumption of anarchy. These different assumptions lead to different research agendas, liberal institutionalists focusing on international environmental regimes, while realists focus on the potential of global environmental change to cause international conflict.

I have also tried to show that while they by and large eschew direct discussion of the causes of environmental change, they nevertheless contain in their work assumptions concerning such causes. These are that environmental change is the result of the overexploitation of commons, of resources owned by no one, and of a set of discrete trends generating such change. In some formulations (Vogler, 1992; Litfin, 1993) these explanations are combined. The use of the former explanation is often inconsistent, in that the way in which Hardin's original formulation emphasising that the lack of property rights over resources is a cause of environmental change is ignored – the 'tragedy of the commons' simply emerges as an explanation for why cooperation is difficult to achieve. But ignoring this part of the implication of using a 'commons' metaphor creates an inconsistency. The second explanation begs the question: what underlies trends such as population or technological change? In later chapters, I will try to illustrate how trends which generate global environmental change are best understood as integral to the reproduction of the major power structures of world politics. In the following chapter I will outline a perspective on global environmental politics which takes this argument seriously. Here I have shown that while some authors have tried to develop such a structural understanding (including Kurian, Bartlett and Malik, 1995; Hempel, 1996; Elliott, 1998), they have only gone half-way, with contradictory results.

3
The 'normal and mundane practices of modernity': Global Power Structures and the Environment[1]

The previous chapter offered a critique of the major theories within IR which address environmental problems and politics. One of the main conclusions was that those theories have only an implicit explanation of how environmental problems emerge, and that the implicit explanation they have is dubious at best. This chapter tries to build a theoretical approach which explicitly incorporates an explanation of the emergence of global environmental problems into the framework, alongside an account of how societies do and should respond to those problems. In the earlier sections, I confine my discussion to how the positions I advance explain the emergence of global environmental change. I discuss responses to such change in the concluding section to this chapter; many of the points developed there are taken up again in Chapter 7.[2]

Green politics and International Relations

Such an approach should perhaps be set in the context of the emerging literature on Green political theory in IR. The focus of Green political theory is usually on two themes: on the role of anthropocentric ethics in reducing the non-human world to being merely of instrumental value to humans; and on the question of limits to growth. Writers in IR have started to engage with Green theory, but this literature is still small.

There is now a well-developed literature on Green political theory (GPT), which gives a useful base for Green ideas about IR. Three major works suggest slightly different ideas about the defining characteristics of Green politics. Eckersley (1992) suggests that the defining characteristic is ecocentrism – the rejection of an anthropocentric world-view which places moral value only on humans in favour of one which places independent value also on ecosystems and all living beings. Goodin (1992) also places ethics at the centre of the Green position, suggesting that a 'Green theory of value' is at the core of Green political theory. His formulation is that for a Green theory of value, the source of value in things 'is the fact that they have a history of having been created by natural processes rather than by artificial human ones' (ibid., p. 27).

Dobson (1990) is the one of these three to have two defining characteristics of Green politics. One is the rejection of anthropocentrism, as outlined by Eckersley. The other, however, is the 'limits to growth' argument about the nature of the environmental crisis. Greens suggest that it is the exponential economic growth experienced during the last two centuries which is the root cause of the current environmental crisis. Thus it is not the belief in an environmental crisis which is defining, but the particular (and unique) understanding which Greens have of the nature of that crisis which makes them distinctive; it is this which, as Dobson, distinguishes Greens from reformist environmentalists.

Dobson's position is the most convincing, in my view. A reduction of the Green position to an ethical stance towards non-human nature, without a set of arguments about *why* the environment is being destroyed by humans, seems to me to lose much of what is central to Green beliefs and practices. In addition, Goodin's formulation is highly problematic, as he posits a notoriously dubious distinction between things which are 'natural' and those which are 'artificial', which cannot be even loosely sustained. Dobson's two principles can be seen to underlie the four principles given in the often-cited Programme of the German Green Party (1983; Mellor, 1992); ecology, non-violence, decentralisation and social justice. As formulated by Dobson, and I would agree, the last three of these can be derived from the first, understood ethically as ecocentrism and empirically as limits to growth (although of course the principles could also be derived independently of these connections).[3]

Ecocentrism[4]

For Eckersley (1992) ecocentrism has a number of central features. Firstly, it involves some empirical claims. These involve a view of the world as composed of interrelations rather than individual entities (1992, p. 49). All beings are fundamentally 'embedded in ecological relationships' (p. 53). Consequently, there is no convincing criteria which can be used to make hard and fast ethical distinctions between humans and non-humans (pp. 49–51).

Secondly, it has an ethical base. Eckersley rejects anthropocentrism on consequentialist grounds, suggesting that it leads to environmentally devastating results (p. 52), but also argues for ecocentrism on deontological grounds. Since there is no convincing reason to make rigid distinctions between humans and the rest of nature, a broad emancipatory project, to which she allies herself, ought to be extended to non-human nature. Ecocentrism is about 'emancipation writ large' (p. 53). All entities are endowed with a relative autonomy, within the ecological relationships in which they are embedded, and therefore humans are not free to dominate the rest of nature.

Ecocentrism therefore has four central ethical features which collectively distinguish it from other possible ethical positions towards the environment.[5] Firstly, it recognises the full range of human interests in the non-human world, not simply the narrow interest in instrumental use of particular parts of nature. Secondly, it recognises the interests of the non-human community. Thirdly, it recognises the interests of future generations of humans and non-humans. Finally it adopts a holistic rather than an atomistic perspective – that is, it values populations, species, ecosystems and the ecosphere as a whole as well as individual organisms (p. 46).

Limits to growth

Although the idea clearly has a long lineage, the immediate impetus for arguments concerning limits to growth came from an influential, controversial and very well-known book published in 1972, *The Limits to Growth* (Meadows *et al.*, 1972). The argument there was that exponential economic and population growth of human societies was producing an interrelated series of crises. This exponential growth was producing a situation where the world was rapidly running out of resources to feed people or to provide raw material for continued industrial growth (exceeding *carrying capacity* and *productive capacity*), and simultaneously exceeding the *absorptive capacity* of the environment to assimilate the waste products of industrial production (Dobson, 1990, p. 15; Meadows *et al.*, 1972). The team of researchers led by Donella Meadows produced their arguments based on computer simulations of the trajectory of industrial societies. They predicted that at current rates of growth, many raw materials would rapidly run out, pollution would quickly exceed the absorptive capacity of the environment, and human societies would experience 'overshoot and collapse' some time before 2100.

The details of their predictions have been fairly easily refuted.[6] However, Greens have taken their central conclusion – that infinite exponential growth is impossible in a finite system – to be a central plank of their position.[7] The fact that specific predictions concerning the timescale within which ecological limits may be breached were wrong does not prevent us believing in the existence of those limits. Dobson (1990, pp. 74–80) suggests there are three arguments which are important here. Firstly, technological solutions will not work – they may postpone the crisis but cannot prevent it occurring at some point. Secondly, the exponential nature of growth means that 'dangers stored up over a relatively long period of time can very suddenly have a catastrophic effect' (ibid., p. 74). Finally, the problems associated with growth are all interrelated. Simply dealing with them issue by issue will mean that there are important knock-on effects from issue to issue – solving one pollution problem alone may simply change the medium through which pollution is carried, not reduce pollution.

From this Greens derive their notions of sustainability. While environmentalism concentrates on 'sustainable development',[8] which presumes the compatibility of growth with responding to environmental problems, Greens reject this. Sustainability explicitly requires stabilising, and in the industrialised countries almost certainly reducing, throughputs of materials and energy (Lee, 1993). This requires wholesale reorganisation of economic systems. Of course many problems exist with the notion of limits to growth. It is often pointed out that if growth is measured by GNP, then this bears no necessary relationship to resource use or pollution levels, since it only measures money flows (for example, by Hayward, 1994, pp. 95–6). This particular critique is perhaps a side-issue however; most ecological critiques of growth focus on the physical flows underpinning the economy rather than directly on GNP. However, it would seem reasonable to assume that were there a serious attempt to minimise resource use and pollution, then this would ultimately have consequences for GNP growth. I develop a discussion of growth again further below, in relation to capitalism.

Green politics and IR theory

It is only recently that a number of writers within International Relations have started to take up the themes developed in Green political theory. Most have simply tried to outline Green political thought for the purposes of people working in International Relations, and in particular to attempt to relate the concerns raised by Greens to those raised within existing perspectives. For example, Hovden (1998), Laferriere (1996), Laferriere and Stoett (1999), Mantle (1999), and Paterson (1996c) all engage in comparisons between Green theory and existing strands of IR theory. Laferriere (1996), for example, contrasts an ecological approach to IR to those of realists and 'critical IR' theories. He makes emancipation the connection between the diverse strands of critical IR theory, and suggests that ecology can provide a 'distinct voice' (p. 73) within critical IR. There are three aspects to the distinctiveness of ecology for Laferriere (p. 74). Firstly, it connects the domination of 'nature' to forms of domination within human societies. Secondly, it provides ways out of the debate in critical theory about foundationalism by remaining critical of the arrogance of the '*certainty* associated with the modern project' (p. 74), while retaining assumptions about nature that provide a basis for redirecting human societies. Thirdly, ecology is able to connect peace to emancipation (pp. 74–5). Laferriere and Stoett (1999) develop the connections made by Laferriere between Green thought and the three strands of IR theory in much greater detail. Hovden (1998) develops similar arguments, but is more careful to distinguish between critical theory (that is, the Frankfurt School) and poststructural IR theory, and argues more forcefully that Green theory has more in common with the former than the latter. Mantle (1999) corrects the absences of feminism from the discussions of Hovden and of Laferriere and Stoett, and persuasively argues that the closest connections which Green theory has to other approaches in IR are to feminism. Helleiner (1996), contrasts a Green perspective with those of liberals, Marxists and economic nationalists in IPE. All suggest that Green IR theory could develop into a distinctive perspective within IR, although most also discuss connections with existing theoretical perspectives, particularly Marxian, critical-theoretic, feminist, and poststructuralist.

The purposes of Kuehls (1996), Doran (1995), Dalby (1998b) and Stewart (1997) are slightly different. They all advance arguments from poststructuralist positions which are also all, at least at a fairly fundamental level, consistent (in their eyes at least) with a Green position. Kuehls' intention is to interrogate Green (or 'ecopolitical') theory as much as that of IR; to argue (among other things) that Green theorists, like IR theorists, remain committed to models of politics based on sovereign notions of political space (see the final section to this chapter). In a similar vein, Dalby (1998b) also questions Green commitments to localism, suggesting instead non-territorial notions of political space, based on networks, like Kuehls' use of the 'rhizomes' of Deleuze and Guattari (see also Chapter 7).

Global power structures and global environmental politics

There is clearly an explanation in the literature on Green political theory and on Green politics and International Relations, of the origins of global environmental change. For Laferriere and Stoett (1999), for example, such causes are political. They are the result of material growth, the use and development of state power, and Newtonian science (1999, p. 4, p. 72). They summarise the ecological crisis as a crisis of 'domination and exploitation' (p. 98). Kuehls similarly locates the origins of global environmental change in accumulation and practices of 'governmentality' (Kuehls, 1996, p. 83; passim).

But the weakness of such arguments is that, as with those of Hempel or Choucri discussed in Chapter 2, they are not structuralist enough. There is a need to understand the structures which produce these ethics and social imperatives. Anthropocentrism does not exist in a social and historical vacuum; rather it has emerged as part of an ideological system underpinning the emergence of modern science, of capitalism, of the modern state, and of specifically modern forms of patriarchy. Likewise, notions of limits to growth need to be supplemented with an understanding that growth is a systemic imperative for capitalist economies, and has been systematically promoted by states both because of their relationship to capitalism, and because of the dynamic of interstate competition and state-building. I am perhaps overemphasising the extent to which this is not part of the understanding of many Green theorists and their interpreters, but nevertheless I think the tendency exists.

My argument is that the politics of global environmental problems should be understood as phenomena internal to the logics of four main, interrelated, power structures of world politics: the state system; capitalism; knowledge; and patriarchy. These power structures are crucially implicated in the generation of environmental problems, and cannot uncritically be used in responding to those problems. The political implications of the global environmental crisis therefore lead to a challenge to these structures themselves, and the normative claims of studying global environmental politics are for a politics of resistance, not of improving the way international treaties and institutions operate. Such an argument, I would argue, is consistent with the themes emphasised in Green political theory. If ecocentric ethics understand the problem of anthropocentrism to be bound up with broader dynamics of domination (patriarchal, statist, scientific) and limits to growth are understood in terms of the political-economic dynamics of capitalist accumulation through which such limits are breached, then this is simply one way of specifying these two themes.

This argument is not entirely original. Within International Relations, a similar argument has been made by Ken Conca (1993). His focus is slightly different, in that it is on whether 'global ecological inter-dependence' is something which may produce structural change in the major structures of world politics (rather than the question of how those structures generate environmental problems). He focuses on capitalism, the states system (with a focus on sovereignty) and moder-nity (taken to mean the hegemonic ideological form which dominates contemporary societies). But his assumption is clearly that those struc-tures are problematic from an environmental point of view. 'The glob-alization of sovereignty, capitalism, and modernity has not been premised on principles of ecological sustainability', he writes (Conca, 1993, p. 321).

Julian Saurin (1994) makes a similar argument. Drawing on Giddens, he uses the concept of modernity (here meaning a broader complex of practices than simply knowledge and ideology as identified by Conca) to refer to the broad set of social structures and practices emerging originally in north-west Europe but gradually becoming globalised over the last several centuries. He places 'technical rationalism' (1994, p. 49) at the core of modernity, but it subsumes many of the features identi-fied here, including capital, bureaucratic state practices and modern scientific knowledge. 'The core claim about modernity', he writes, 'is that large-scale and systematic degradation occurs from its ordinary and standard practices' (1994, p. 46).[9] While the practices producing environmental problems are perhaps mundane, they are very much the '*normal and mundane practices of modernity*' (Saurin, 1994, p. 62).

However, these arguments are particularly poorly represented in studies of environmental problems within IR. That field is dominated by the work of writers in traditions discussed in Chapter 2. Arguments such as those put forward by Conca, Saurin, or many feminists (such as Seager, 1993) are still marginal within the study of global environmen-tal politics. In what follows I focus primarily on how the structures I identify systemically generate global environmental change. I turn later to questions of political responses.

It should hopefully be reasonably clear that when I use the term structure I do not intend this to mean that such structures are immutable, inexorable. I regard such structures as conceptual contructs which help us understand the complexities of the social world. I regard them as historically evolving, and as undergoing transformations.

Nevertheless, it is useful to regard them as structures in the sense that they *tend* to reproduce their basic principles through the practices of actors operating within them, and thus tend to reproduce the way in which they confer power on actors differentially. They thus act on a variety of actors (individuals, firms, states, a variety of collectivities) both in terms of constraints on action, and also in terms of producing identities and thus practices. Understanding the dynamics of these structures is therefore of key importance in establishing the possibilities for agency in responding to global environmental change.

Robert Cox gives a good account of how I understand the usage of the term structure, and is worth quoting at length:

> Some authors have used 'structure' to mean innate ideas or patterns of relationship that exist independently of people; they think of people as merely bearers of structures. No such meaning is intended here. There is, of course, a sense in which structures are prior to *individuals* in that children are born into societies replete with established and accepted social practices. However these practices, whether taking the forms of language, legal systems, production organization, or political institutions, are the creation of collective human activity. Historical structures, as the term is used in this book, mean persistent social practices made by collective human activity and transformed by collective human activity.
>
> (Cox, 1987, p. 4)

Cox focuses primarily on the production relations of capitalism while I try to cast the net more broadly. But the understanding of structure remains broadly the same as that adopted by Cox and other neo-Gramscians (for instance, Gill, 1993).

The states system

There is of course a minefield involved in defining the states system (or the other structures I discuss) and I will inevitably privilege some views over others. Nevertheless, some specification of that system is necessary.

As it has emerged historically, I would characterise the modern states system in terms of the consolidation of institutional complexes of power around territorially defined states. Such consolidation has involved state elites in concentrating control over 'their' territories, and in producing both the land as territory and the people occupying that land as citizens (see, for instance, Kuehls, 1996, chs 2 and 3). At the same time it has engaged them in acts of legitimation, both internal and external. Internally it has involved building constituencies of support for their rule, and identities which promote 'national' cohesion – 'imagined communities' (Anderson, 1983). Externally, it has involved the construction of sets of rules by which separate states can interact with each other in an orderly way. Perhaps the most fundamental of these has become the institution of sovereignty, whereby states have produced each other as formally independent, but it also involves many other rules and conventions, as for example in Bull's account (1977, p. 74 and chs 5–9) of the institutions of international society: the balance of power, international law, rules of diplomacy, war, and the special position of great powers.[10]

In such a view, war-making has been a necessary component of state-making (Tilly, 1985). It is not just that war has been a means for states to increase their power *vis-à-vis* other states. It has also been a means through which a population has been made into a 'nation'; a disciplined group of people with a shared sense of identity. And this shared sense of identity has been constructed through opposition to others. Fighting others has been a means though which these national identities have been produced.

Such a view of the states system is by no means incompatible with a view that international cooperation, such as over environmental change, can be extensive. It is not simply saying that the international system is inevitably conflictual. Rather, the point here is that this system has certain internal dynamics which have helped to produce environmental change. Although what follows is inevitably an oversimplification, these dynamics can be summarised in this way.

Firstly, state-building, both to intensify control over territories internally, and to wage war against other states more effectively, has meant that state elites have systematically promoted accumulation. They have needed primarily to extract surplus from populations more efficiently, in order to generate resources for such state-building. This has helped both to create a surplus which has then been used in part for investment in order to promote further accumulation (the development of military technology also involved here has had spin-offs for non-military uses) and to create incentives to increase incomes in order to be able to pay the new taxes without becoming worse off. This was in particular helped by shifts in forms of taxation from being in kind (tithes, and so on) to being in cash (taxes), meaning that people had to participate in a cash economy to be able to pay their taxes. Through this means alone, states helped to expand the scope of a market-based economy. Accumulation, and particularly resource-intensive accumulation (states having interests in heavy use of resources needed for warfare – timber for ships, iron/steel, coal, etc.), has therefore been an integral part of the operation of the states system. Finally, as populations were increasingly mobilised to serve state requirements, states had to legitimise themselves to those populations. This is perhaps part of the explanation of the origins of welfare states (Mann, 1986) but in particular it becomes an additional dynamic underpinning the accumulation imperative for states – increased consumption becomes a way of legitimising state rule. As the following section outlines in more detail, states also promote accumulation due to their role in reproducing capitalist social relations.

Secondly, the military competition which has been present throughout the history of the states system has itself generated environmental change. In part, this is because the environment has been an instrument and a casualty of warfare itself, as strategists have used and abused ecosystems to give themselves military advantage (Finger, 1991; Westing, 1986). But, in addition, military-intensive development has generated a particularly heavy use of highly toxic chemicals, explosives, and so on, which themselves produce environmental degradation, even in peacetime (see Seager, 1993, pp. 14–69 for example).

Thirdly, the states system has facilitated various acts of what might be called environmental displacement. The existence of borders, combined with a (globalising) capitalist economy has meant that environmental degradation can be exported. International inequalities can be used so that rich countries can exploit the desires of poorer countries for export earnings or investment by exporting either pollution directly (as in toxic wastes, see Puckett, 1992, 1994; Wynne, 1989; Clapp, 1994) or dirty industries which then pollute elsewhere (see e.g. Seager, 1993, pp. 154–6). Displacement can also occur in responses to environmental change, where the interstate system allows state elites to direct attention and blame on to others for causing environmental change (Hay, 1994).

Finally, at a more philosophical level, the modern state both represents, and has sedimented in other areas of social life, abstracted notions of hierarchy and domination in their purest forms. Bookchin (1980; 1982), for example, suggests that the state is the ultimate hierarchical institution which consolidates all other hierarchical institutions. Such institutions of domination simultaneously involve the domination of some humans by others and the domination of non-humans by human societies. The political institutions of rule cannot therefore be disconnected from the 'domination of nature'. Furthermore, forms of state rule under modernity have become progressively rationalised and bureaucratised. Such abstract systems of rule rely on similarly abstract forms of knowledge and ways of knowing, which feminists and many ecologists suggest are deeply problematic in ecological terms. I will return to this topic.

Through these dynamics, it is at least historically the case that states and the states system have generated environmental change as a product of their internal operation. Through the promotion of accumulation, military competition, practices of ecological displacement, and rationalised rule, states systemically produce environmental degradation. It is of course a stronger argument to say that they cannot be reformed so that they do not produce such environmental change or degradation in a systematic fashion. Much of this sort of a debate ends up turning on a semantic discussion of when is a state (system) not a state (system), which is not particularly useful. I turn to questions of political change in Chapter 7. For present purposes, it is enough perhaps to demonstrate that the operations of states have historically produced environmental change, and that this cannot plausibly be thought of as an 'accident'.

Thus, for example, Litfin is mistaken when she writes:

> The traditional goals of the modern state – 'to defend borders and promote industrial development' [Princen, Finger and Manno, 1995, p. 50] – are arguably in friction with the quest for ecological integrity. But no a priori reason exists for saying that environmental protection cannot become one of the state's primary objectives, and there is evidence that it is doing just that.
>
> (1997, p. 195)

By contrast, there are very good *a priori* reasons for claiming that 'environmental protection cannot become one of the state's primary objectives', and to the extent that 'it is doing just that', such a goal is in fundamental conflict with other state imperatives. Unlike those imperatives, however, sustainability cannot become a structural requirement for the reproduction of state rule, and thus the system must be transformed.

Capitalism

As with the states system, reaching a working definition of capitalism is far from an uncontroversial exercise. I start however with what I take to be a fairly conventional account derived from (although not used exclusively by) Marxism. Capitalism is a social system which is based primarily on the commodification of labour; that is, that human labour is treated like any other commodity, and has a price which can be traded freely in the market. It is the efficiency of this form of extraction of surplus value from labour which has generated the system's extraordinary productive potential (rather than the efficiency of a 'free market' as argued by liberal economists).

The ways in which capitalism has been systemically ecologically damaging can be outlined along (at least) four general dimensions. Firstly, the productive potential of capitalist social relations (wage labour) combined with the competition facing capitalists in the marketplace, and the technological dynamism produced by these two features, mean that the capitalist economic system *requires* growth in order to survive. Perhaps the simplest expression of this is that within modern capitalist economies a lack of growth is *the* definition of crisis (recession). Growth on the scale of the economy as a whole is the corollary of the need for firms in a competitive situation to maximise profits. As firms make those profits and recirculate them in terms of investment or consumption, new investments mean greater productivity and further increases in production and consumption. If the system as a whole were unable to grow, then individual firms would eventually run out of profitable investments.

In addition, because of the necessity of economic growth for capitalism to survive, those organising such growth, defined generally as capital, gain a great deal of power with respect to state decision-making. In this context, therefore, the state's fundamental purpose is to secure the conditions under which capital accumulation (economic growth) can proceed smoothly. This involves the state in securing collective goods (such as defence and the rule of law) so that growth can occur, mediating social conflicts produced by class relations, appearing as neutral so that class domination is obscured. (This is in addition to any growth dynamic identified earlier with regard to the states system.) Therefore, those who own the means of production gain a structurally powerful position within states. They come to have veto power in relation to state policies.

As a consequence, even if one were unpersuaded by the abstract limits-to-growth argument, then whenever the formulation of environmental policies hurt the capacity for businesses to make adequate profits, these policies will come up against opposition from the most powerful group in society. The growth dynamic of capitalism provides a powerful constraint against responding effectively to environmental change. This means that states face inevitable legitimation crises in the face of environmental crises; crises imposed by a contradiction between the structural necessity to promote capital accumulation, and the necessity to legitimate state practices, which often requires making some attempt to resolve environmental problems (Hay, 1994).

The limits-to-growth argument should be understood in this context. Most Greens tend simply to assert the existence of such limits, and argue that we need a change of social attitudes, voluntary simplicity, and so on. The analysis here suggests that basic structural constraints affect the prospects for moving towards such goals, and the latter require much more than simply changes in lifestyles and government policies. Such a goal requires changing the basic forms of social relations, away from those which systemically generate and require growth. This should not, however, be taken as an argument wholly for an ecosocialist position on the question of 'industrialism vs capitalism'. Atkinson (1991, p. 5) has a resolution of this I find persuasive: 'In practice there is no fundamental contradiction between these views. If we are to de-escalate our ecological crisis then it will be necessary to restructure productive industry along the lines envisaged by the Greens. But it is also true that any headway in this direction will be made over the dead body of capitalism; the very soul of capitalism is the requirement for economic growth ... '.

A second ecologically damaging dynamic of capitalism concerns the notion of commodification. One of the main secular trends of capitalism has been to commodify increasing amounts of the world. In other words, more and more things become commodities, things produced for sale in markets. One ecological consequence of this, as has often been pointed out, is that 'nature' has become 'natural resources'. Capitalist development has involved humans breaking down the complex wholes of ecosystems, and so on, into their constituent parts in search of economically profitable resources for production. Commodification thereby renders the world as both an object, and more precisely as a set of objects which are treated independently of each other. Thus the interactions between different parts of ecosystems, necessary for the continued functioning of the biosphere, are obscured from view by the rationalising imperative of capitalist production.

Thirdly, capitalism necessarily generates environmental problems because of the way in which firms must subordinate all other concerns to the primary goal of profit-maximisation. Profit-maximisation is particular to the nature of capitalist markets (as opposed to, for example, guild-organised markets which were prevalent in medieval

Europe) because of the way in which restrictions on competition were progressively stripped away as capitalism emerged; firms face sufficient insecurity (no regulations exist to guarantee their existence) and therefore they need to maximise profits in order simply to maximise their chance of continued existence. As a result, other concerns, such as the sustainability of their practices, have to be subordinated to profit maximisation.[11]

This argument is in many ways consistent with a focus on externalities as generators of environmental problems. For liberal economists, the ultimate origin of any environmental problem is the lack of property rights assigned to the resource, ecosystem, etc., which is degraded. A consequence of this is that goods traded in markets do not have the value of such resources included in them – the value of (for example) a stable climate is an *externality* in relation to the price of coal. But for this critical reading, the fact that firms do not have to take environmental costs into account is still a large part of the explanation of why environmental problems emerge. However, this perspective might also suggest that in practice firms are likely to resist attempts to incorporate those environmental costs, as this could affect their competitiveness.[12] This would be especially the case since the emergence of a genuinely global economy, whereby accelerated and deregulated capital movements enhance the exit options and hence structural power of those firms resisting the development of environmental regulations.

A final ecologically damaging feature of capitalism concerns inequality. Many (not only Marxists) would suggest that capitalism necessarily generates increasing inequality, particularly on a global scale. This would be the position adopted, for example, by most Marxist writers on global politics, as well as dependency theorists and many Greens. Historically, the practices of imperialist states have very clearly impoverished large parts of the world. But this argument is slightly stronger, that the continuation of capitalist economies depends on the creation of inequality. Marx's explanation for economic crises in capitalism centred on an 'underconsumption' hypothesis – that the tendency for individual capitalists to pay workers only subsistence wages meant that the capacity of capitalists collectively to realise value was compromised since there would not be enough effective demand for the goods produced. This is also part of his explanation for why capitalism could not survive as a social system. One common explanation for the inaccuracy of his predictions of capitalism's demise is the emergence of the welfare state which cushioned the 'immiseration of the masses', but perhaps more importantly the emergence of employers who strategically raised wages to try to get round this problem. Henry Ford's five-dollar day is often cited as important here.

But perhaps this inequality was simply displaced to global levels. The industrialisation of the West was made possible by the extraction of cheap raw materials using cheap labour from colonies, and the continued growth of the world economy is still dependent on the ability to have sharply differing standards of living across the globe (and arguably within particular societies as well), for example, to be able to exploit cheap labour for particular parts of production.

Such global inequalities have clearly produced ecological consequences. The clearest examples of these concern the relationship between the debt crisis, structural adjustment, and deforestation. The export dependence of many developing countries has meant that once they incurred large debts to fund industrialisation programmes in the late 1970s, and were unable to repay them following the oil price-hike of 1979 and the interest-rate rises of the early 1980s, they turned to increasing exports of their natural resources to increase their earnings to pay off these debts. Such a process was intensified by IMF- and World Bank-imposed policy changes known collectively as structural adjustment. Even without the debt crisis and its political responses, however, the organisation of the global economy has meant that many countries depend heavily on using their natural resources to be able to earn foreign currency and participate in the global economy. A particularly stark example of this concerns toxic wastes, where what Lawrence Summers, a World Bank (and later Clinton administration) official, termed the 'underpolluted' nature of many developing countries in economic terms has meant that their poverty has induced them to take in dangerous wastes which richer countries export. Within capitalism, such a development is not just inevitable, but rational. As Summers pointed out, 'The economic logic behind dumping a load of toxic waste in the lowest wage country is impeccable' (as cited in Karliner, 1997, p. 148).

Combined, such arguments suggest that a capitalist society necessarily generates global environmental change. Its basic productive drive leads it to generate an ever-intensified throughput of resources. It also continually must increase the range of things brought into the realm of market exchanges in order to generate new sources of accumulation, and treats such things as commodities, as objects for human instrumental use. As firms are forced into 'cut-throat', 'grow or die' strategies, profit margins become the ultimate source of value and all other values, including ecological ones, must be subordinated to that imperative.

Moreover, the accumulation process depends on exploiting and intensi-fying global differentials in income, generating incentives for the poor to rely on particularly resource-intensive forms of development in an attempt to 'catch up'.

Knowledge and power

The third structure is that surrounding the production and distribution of knowledge. Mainstream IR views focus on how scientific knowledge, and 'rational management', are essential for successful responses to global environmental problems. This would be in line with 'common-sense' arguments that good scientific knowledge is necessary both to be able to identify environmental problems, but also to provide the tools to respond effectively to them. Thus a prevalent argument is that international cooperation on environmental problems depends on suf-ficient availability of scientific information to be able to assess the rationality and effectiveness of various strategies, but also on the exis-tence of an epistemic consensus among the relevant scientific experts (see, for example, Andresen and Ostreng, 1989; in its 'epistemic communities' variant, see Haas, 1990a).

Again, however, many critics interpret modern scientific rationality and scientific institutions as underlying structural causes of environ-mental problems. There are two aspects to this argument. Firstly, mod-ern science was founded on the dualistic assumption of human separation from and domination over the rest of the natural world, and in fact for many scientists its purpose has been precisely to further this separation and domination. Many writers suggest that this has led to anti-ecological attitudes and practices. This is because the rest of the natural world has been reduced to an object for human instrumental use, whereas conceiving it as an end in itself would probably produce less ecologically damaging behaviour. It is also because of the way in which science (or at least, dominant traditions within science) has adopted a reductionist methodology, where phenomena are reduced to their constituent parts, and analysed as individuals. Science has there-fore been less well focused (perhaps only until recently) on the interac-tions between things; yet it is primarily in these interactions that environmental problems emerge. This account is developed by, for example, Plumwood (1993), Merchant (1980) White Jr (1967) and Smith (1996). On the emergence of more ecological, holistic, approaches within science, see in particular Worster (1994).

As scientific endeavour has expanded with modern societies, the nature of the hazards created by modern technologies is such that the experimental logic which could be used to generate 'proof' of environmental degradation (as assumed by those advocating science as the solution to the environmental crisis) fails to achieve its potential (Beck, 1995). It is this which Beck terms 'organized irresponsibility'; the systemic production of large-scale technological hazards occurs in a manner which makes clear connections of causality, responsibility and proof impossible to ascertain (ibid.).

The second aspect is that the emergence of science, particularly in combination with the emergence of capitalism and the modern state (which have gone alongside each other), and the transformation of patriarchy produced in part by modern science, has transferred legitimacy concerning knowledge about environmental problems to particular elites, whereas without science they would have been more evenly distributed. Science has become a way in which control over environments has been taken away from individuals or communities and given to experts, who increasingly live away from the environments which they are charged with 'managing', and thus have no personal interest in whether the management of those environments is sustainable, or whether it meets the needs of those who do depend on it. But if successful responses to environmental problems rely on those who depend on resources being able to control how they are used, then at the very least the particular organisation of modern science (being elitist rather than democratic) is problematic from an environmental point of view (Ecologist, 1993, pp. 67–9, 183–6; Banuri and Apffel-Marglin, 1993; Beck, 1995, ch. 7; Gorz, 1994).[13]

Patriarchy

The fourth structure is that of patriarchal power. Here, the argument is slightly different, since no one (explicitly) argues that patriarchy is good for the environment. It is more that mainstream views would be blind to the impact of patriarchal forms of power on global environmental problems. However, a full understanding of the politics of these problems must recognise the gendered distribution of power, resources and identity.

This distribution both generates global environmental problems and constrains the ability of proposals to respond to environmental problems. Three strands (at least) to an argument connecting patriarchy to the generation of environmental change can perhaps be identified. Firstly, central to modern masculine identity is individualism – the notion that humans are fundamentally independent individuals, and should strive to avoid dependence on others. Modern political ideas of freedom are heavily dependent on this form of identity. Feminists have shown how this idea has in modern times applied only to men, but is in any case false empirically. The independence of men depends in fact on the subordination of women, that women are doing the 'reproductive labour' of providing food and clothing and looking after the home, as well as reproducing, and bringing up children. As Plumwood refers to it, such dependence and the work of women in producing the basis for mens' 'independent' lives is 'backgrounded' (Plumwood, 1993).

The effect, however, of individualist ideology on environmental problems has been that as individual freedom has been made paramount, it has become more difficult to question the knock-on effects of people's actions (in economist's terms the 'externalities'). Therefore proposals to solve environmental problems, which involve curbing this individual freedom, are more difficult to promote. Feminists suggest that a marginalised rationality, which is not individual but communal, and which has traditionally been associated throughout patriarchy with women, is more appropriate for dealing with environmental problems.

Secondly, and of course linked to masculinist individualism, is the notion of instrumental rationality. Feminists have shown how the emergence of modern science in the sixteenth century can be implicated in producing environmental problems. Carolyn Merchant in particular shows this in *The Death of Nature* (1980). The emergence of modern science, with people like Francis Bacon, involved a shift from thinking about the natural world as something on which humans depended, which may at times be difficult, but something over which humans had little control, towards an idea that nature was something which humans could and should 'master' or dominate. Francis Bacon was very explicit about this. What emerged was the idea that nature only really exists for human use. This instrumental rationality – that it is rational and right to use things as and when they are useful to you – has as its consequence that there should be no moral constraints on this instrumental use. But this has arguably meant that it has become much easier to pollute the rest of the world, and to overexploit nature, since it is now seen merely as 'natural resources' (just there for human use).

This emergence of instrumental rationality was also highly gendered. Women in Bacon's language were not fully rational, and therefore were considered as part of nature, not part of humanity. They were therefore to be used (and abused) by men just like animals, plants and rocks.

This is very explicit in his writings, and is a prevalent idea in Western thought – which identifies men with culture, and women with nature. It is this, perhaps more than anything, which is at the root of claims by feminists that the domination of women by men and the domination of nature by humans are closely linked. Indeed, for some, the domination of women by men is often regarded as having been one of the earliest forms of human domination, and one which enabled later forms of domination to emerge, in particular the domination of (the rest of) nature by humans (e.g. Bookchin, 1982).

The route out, for feminists (as also for many environmentalists), is to stop thinking about nature as simply 'natural resources', simply something there for human use. Also, this involves rethinking science, moving away from 'reductionist' and 'dualist' science (see earlier), and towards 'holistic' science, which sees the world as a whole rather than merely the sum of its parts. Then, it is claimed, it will be easier to identify the connections between species, ecosystems, etc., and therefore the effects which human interventions in the environment might have.

The interaction of these two features of (modern) patriarchal societies can be used to analyse Hardin's (in)famous 'tragedy of the commons', already mentioned in different contexts. In this context, it is clear that the herds*men* are following individualist, and instrumental rationality. By thinking only about themselves, and aiming merely to use the commons rather than preserve it, they overgraze. While it is of course simplistic to suggest that if they had been herds*women*, these rationalities would have been less likely to prevail and overgrazing to occur, the point that the modes of rationality which prevail in Hardin's metaphor are masculinist and specific to (modern, capitalist) patriarchal societies remains (see Mellor, 1992, pp. 232–3, for a feminist critique of Hardin's thesis).

The third plank of a feminist argument about the environment is a power-based one. It suggests that male domination over women means that the effects of actions which damage the environment can be divided up so that the polluters (mostly white, affluent, Western men) do not suffer the consequences – these are felt primarily by women, non-Westerners, and ethnic minorities in the West.[14] While one effect of this power is to facilitate the generation of environmental change, it can most clearly be seen in the way that patriarchal power remains a significant obstacle to responding successfully to environmental problems. Male power in many societies still means that men are able to displace the effects of environmental problems. For example, it is well documented that within agriculture in many developing countries, men arrogate commercial agriculture to themselves, making sure that they work on the most fertile land, and that they can participate in the cash economy. Women are predominantly left with responsibility for subsistence agriculture, on the poorer land, as well as with responsibility for other subsistence activities such as collection of water and wood. Thus they bear the brunt of environmental problems such as desertification and deforestation, having to travel progressively longer distances to meet subsistence needs (Sontheimer, 1991; Dankelman and Davidson, 1988).

The operation of modern patriarchy is, of course, inextricably bound up with the emergence of modern science, capitalism and the state system. Modern science, at its inception with people such as Bacon and Descartes, had the defining of women as nature and therefore as objects, as one of its core projects (e.g. Merchant, 1980; Plumwood, 1993). The emergence of the contract as one of the legal foundations of capitalism and the state in the seventeenth century also had redefining the subordination of women as one of its core aims (Pateman, 1988). Capitalism continues to be dependent on super-exploitation of women's labour in reproducing the labour power of male workers, as well as in child-rearing (Mellor, 1992, pp. 208–12; Henderson, 1978; Mies, 1986; Waring, 1988).

Development and the production of global environmental change

These structures and their effects clearly interact with each other. Some interactions were intimated above, although these connections have not been developed. The argument here is that these structures interact primarily in mutually supportive ways. They reinforce each other in ways which mean that those with power within them are able to maintain their power, and also in ways which exacerbate the effect of each on global environmental politics. Indeed, in a pure sense, it is perhaps misleading to say that they interact with each other, since it would be more accurate to say that they are mutually constitutive of each other – they are each other's condition of possibility. In other words, it is an abstraction to talk of the states system without reference to its patriarchal, capitalist, scientised form. However, it is a necessary abstraction which for our purposes here we have to live with.

The closest conception to the interpretation advanced here is the position developed by Alan Carter (1993). Carter outlines an 'environmentally hazardous dynamic'. This is one where:

a centralised, pseudo-representative, quasi-democratic state stabilizes competitive, inegalitarian economic relations that develop 'non-convivial', environmentally damaging 'hard' technologies, whose productivity supports the (nationalistic and militaristic) coercive forces that empower the state.

(Carter, 1993, p. 45)

This scenario necessarily produces a high degree of production of military weapons, which is environmentally problematic, but it also 'demands ever-increasing productivity, entailing a high consumption of resources and an equally high output of pollution' (Carter, 1993, p. 46).

In other words, it is the network of various power structures which are mutually supportive of each other, and which necessarily produce environmental crises through their interaction.[15] Similar arguments have been made by a whole host of Green writers, as well as by many feminists and some Marxists.[16] Carter's formulation does not specify the patriarchal nature of these power structures, but remains a useful starting-point for analysis.

However, a good case can be made that patriarchy should be treated as the core power structure. Historically speaking, the other three are specifically *modern* structures, whereas patriarchy has a longer history. It is also clear that the projects which produced science, the state and capitalism were specifically masculinist and patriarchal in character. As Mellor emphasises, we should 'see the primary role of patriarchy in creating the modern scientific, industrial and military systems that are threatening the planet' (1992, p. 53).

A useful way of seeing their interaction is through the lenses of those critical writers on environmental politics who have focused on development as a central problematic. Development as they understand it can, in the light of the argument outlined above, be seen as a discourse through which these power structures are reproduced. It also can help to show empirically how they interact to produce their ecologically and socially problematic outcomes.

Against development[17]

Writers such as Wolfgang Sachs do not believe the term development can be retrieved.[18] They are highly critical of the term 'sustainable development', in widespread use in environmentalist circles, suggesting that this merely serves to make it easier for ruling elites to coopt environmentalism. Sachs writes, illustrating this argument well:

The walls of the Tokyo subway used to be plastered with advertising posters. The authorities, aware of Japan's shortage of wood pulp, searched for ways to reduce this wastage of paper. They quickly found an 'environmental solution'; they mounted video screens on the walls and these now continuously bombard passengers with commercials – paper problem solved.

(Sachs, 1993a, p. 3)

In other words, elites manage to deal with environmental problems discreetly, while in practice ongoing development undermines any ameliorative effect which a particular response, such as changing the medium of advertising on the underground, may have.

One of the reasons why the 'global ecology' writers object to development is a belief in limits to growth arguments that were abandoned by much of the environmental movement during the 1980s, largely for tactical reasons, in favour of 'sustainable development' or 'ecological modernisation'. Implicit throughout their work is a need to accept the limits imposed by a finite planet, an acceptance which is ignored by the planet's managers and mainstream environmentalists. 'In the eyes of the developmentalists, the 'limits to growth' did not call for abandoning the race, but for changing the running technique', writes Sachs (1993a, p. 10). They are also sceptical of the idea that it is possible to decouple the concept of development from that of growth. While many environmentalists (e.g. Daly, 1990; Ekins, 1993) try to distinguish the two by stating, in Daly's words, that 'growth is quantitative increase in physical scale while development is qualitative improvement or unfolding of potentialities' (Daly, 1990, in Ekins, 1993, p. 94), others would suggest that in practice it is impossible to make such neat distinctions. For the practitioners of sustainable development, 'sustainable development' and 'sustainable growth' have usually been conflated, and certainly the Brundtland Commission regarded a new era of economic growth as essential for sustainable development (WCED, 1987).[19]

However, there are a number of more nuanced arguments which 'global ecology' writers make. While accepting limits in principle, they would be critical of the scientistic fashion in which Meadows *et al.* demonstrated limits – that the computer modelling approach would itself lead easily to a 'global environmental management' form of response which entrenched the power of elites. This was of course one critique of the *Limits to Growth* in the 1970s, that it was too technocratic. They would also agree with another significant criticism of the *Limits to Growth*, for example by Cole *et al.* (1973), that their models had no social content. The social effects of growth, and the social context of developing sustainable societies, are crucial for these writers.

The Ecologist (1993) suggests that one of the central features of development is enclosure, or the turning of common spaces into private property. This was central to modernising agriculture in England before the industrial revolution, and they suggest it is a central part of development practice throughout the world at the present. It is important to development because it is an act of appropriation which makes commodity production possible. Commons were organised largely (but not exclusively) outside the market, making efficient accumulation difficult. Enclosure makes this possible. However, one effect of enclosure is to take decision-making away from those who depend on local resources, which in turn makes environmental degradation more likely, as well as being socially divisive. This argument is closely tied to the argument in favour of the commons, explored below.

As a consequence of enclosure, access to resources is denied, which concentrates resources and power in the hands of fewer people. Development is thus necessarily inegalitarian, since it depends on continuous appropriation. Inequality has been one of the central ideological arguments governments have often made for economic growth; that under conditions of inequality, growth enables the worst-off to get better-off even while inequality is not reduced.[20] An anti-ecological dynamic is therefore built into development. This also illustrates how the global-ecology writers make close links between the damaging human effects of development and the damaging ecological effects of development.

Development therefore entrenches the power of the already powerful. This can be seen on the global level – in the global economy in which the North dominates – and can insulate itself from (many) socio-ecological effects of development, such as through exporting dirty industries to developing countries. It can also be seen at the micro-level, for example in the 'Green revolution' in the 1970s, which concentrated power and land in the hands of the rich farmers, at the expense of the poor who could not afford the fertilisers and pesticides to support the new strains of crops (e.g. Trainer, 1985, pp. 139–41; George, 1977).

A central part of this concentration of power is to do with knowledge. The appropriation of spaces previously held in common empowers 'experts' and denies indigenous knowledges as it transforms those spaces into objects for commodity production. This means that the techniques involved in attempts to manage those spaces are turned over to scientists, and other development experts (The Ecologist, 1993, pp. 67–70; see also Gorz, 1994).

This involves privileging Western technology and knowledge over non-Western knowledges. Thus 'technology transfer' becomes central to solving environmental problems – the idea that 'advanced' Western technology is needed to help developing countries develop in an 'environmentally friendly' way. McCully (1991) provides a compelling critique of technology transfer regarding climate change, showing how past attempts at technology transfer, through development aid, have reproduced the problems associated with development outlined above.

As mentioned above, development necessarily creates commodity production where previously it had not prevailed. This is closely linked to the emergence of instrumental rationality and individualism, which, as mentioned above, has turned 'nature' into 'natural resources', to be plundered by humans (also Shiva, 1988). Development therefore brings about an ideological shift of world-view, a major part of which is towards seeing the environment purely in human-instrumental terms.

Closely allied to this is the idea that development progressively 'rationalises' the natural world. It turns it into a set of countable species, some of which are useful (to be preserved) some of which are not (to be destroyed if in the way of progress). This way of seeing nature has historically reduced biological diversity, and arguably must do so.

The global ecology writers therefore present a powerful set of arguments as to how development is inherently anti-ecological. This is not only because of abstract arguments concerning limits to growth, but because they show in a more subtle fashion how development actually undermines sustainable practices. It takes control over resources away from those living sustainably in order to organise commodity production, it empowers experts with knowledges based on instrumental reason, it increases inequality which produces social conflicts, and so on. At the same time as it is anti-ecological, development as understood by Sachs and others also brings about capital accumulation, intensification of state power with respect to local communities, 'scientifically rational management', and the reorganisation of patriarchal forms of power.

Conclusions

If it is the case that the structures and forms of power prevalent in modern society are anti-ecological, then what sorts of political responses are consistent with this argument? This question can be broken down into two. Firstly, what sorts of political structures might not systematically produce global environmental change? Secondly, how might we envisage transitions from present systems to such sustainable ones? Who might be the agents of such social and political change, and what should they do? I will offer some responses to these questions, although the second will be treated more fully in Chapter 7.

It is perhaps worth starting with the arguments in Green political theory concerning what sorts of political responses are required by the ecological crisis. Drawing only on the ecocentric aspect of a Green position, Eckersley develops a political argument which is statist in orientation. Although she does not adopt the position of the 'eco-authoritarians' such as Ophuls (1977), Hardin (1974) or Heilbroner (1974), she suggests, in direct contradiction to the ecoanarchism which is prevalent in Green political thought, that the modern state is a necessary political institution from a Green point of view. She suggests that ecocentrism requires that we both decentralise power down within the state, but also centralise power up to the regional and global levels.

For Eckersley, then, new forms of global political structures are required from an ecocentric point of view. This is necessary in order to protect nature. Arguing against the anarchist interpretation of Green politics (see below) she maintains that a 'multitiered' political system, with dispersal of power both down to local communities and up to the regional and global levels is the approach which is most consistent with ecocentrism (1992, p. 144, 175, 178). If all power is decentralised, she suggests, there will be no mechanisms to coordinate responses to regional or global environmental problems, or to redistribute resources from rich to poor regions of the world.[21] Her argument is premised on ecocentric ethics and the priority to protect the rest of nature, the social justice consequences of ecocentrism, and the urgency of the ecological crisis. Arguing against ecoanarchists, she suggests that:

> in view of the urgency and ubiquity of the ecological crisis, ultimately only a supraregional perspective and multilateral action by nation States can bring about the kind of dramatic changes necessary to save the 'global commons'... .
>
> (Eckersley, 1992, p. 174)

Her arguments elsewhere are also premised on the urgency of the ecological crisis. 'Indeed, the urgency of the ecological crisis is such that we cannot afford *not* to "march through" and reform the institutions of liberal parliamentary democracy ... and employ the resources ... of the State to promote national and international action' (ibid., p. 154).

This position could be developed within a conventional perspective on IR (such as liberal institutionalism), to look at the character of a wide variety of interstate treaties and practices. The most obvious would be those regarding biodiversity, acid rain or climate change. But it could also be developed for global economic institutions such as the World Bank, or the military practices of states. A broad critique of the major global institutions from an ecocentric position could fairly easily be established, especially considering the very different ethical basis underlying this position in contrast to that which informs international treaties and other international practices. This critique would show how the main international practices are based on an anthropocentric ethic which puts human material interests first, and disregards that of ecosystems or other species. This is even the case for environmental treaties. For example, while ostensibly about protecting biodiversity, the substance of the Biodiversity Convention signed in 1992 is primarily couched in terms of protecting the gene pool for the biotechnology industry (Chatterjee and Finger, 1994, pp. 41–3; Kothari, 1992). And the objective of the Climate Change Convention, also signed in 1992, while stating that the aim is to 'prevent dangerous anthropogenic interference with the climate system', which could have an ecocentric interpretation, quickly goes on to say that this is 'to ensure that food production is not threatened and to enable economic development to proceed in a sustainable manner' (United Nations, 1992, Article 2). As a consequence the implications of ecocentric ethics are limited to a critique of the content of international practices, rather than the structure of international relations.

But this interpretation of ecocentrism advanced by Eckersley is challengeable. Ecocentrism is in itself politically indeterminate. It can have many variants, ranging from anarchist to authoritarian, with Eckersley's social democratic version in the middle of the continuum.

The predominant alternative interpretation within Green thought suggests that it is the emergence of modern modes of thought which is the problem from an ecocentric point of view. The rationality inherent in modern Western science is an instrumental one, where the domination of the rest of nature (and of women by men) and its use for human instrumental purposes have historically at least been integral to the scientific project on which industrial capitalism and the modern state are built. In this reading, environmental ethics are given a historical specificity and material base – the emergence of modern forms of anthropocentrism are located in the emergence of modernity in all its aspects.

This interpretation argues therefore that since modern science is inextricably bound up with other modern institutions such as capitalism, the nation-state and modern forms of patriarchy, it is inappropriate to respond by developing those institutions further, centralising power through the development of global and regional institutions. Such a response will further entrench instrumental rationality which will undermine the possibility for developing an ecocentric ethic. An ecocentric position therefore leads to arguments for scaling down human communities, and in particular for challenging trends towards globalisation and homogenisation, since it is only by celebrating diversity that it will be possible to create spaces for ecocentric ethics to emerge. This argument is developed by the 'global ecology' writers outlined above.

The implications for global politics of an acceptance of limits to growth are clearly considerable. O'Riordan (1981, pp. 303–7; also Dobson, 1990, pp. 82–3) presents a useful typology of positions which emerge from the limits-to-growth version of sustainability which Greens adopt. The first is very similar to that outlined by Eckersley above – that the nation-state is both too big and too small to deal effectively with sustainability, and new regional and global structures (alongside decentralisation within the state) are needed to coordinate effective responses.

A second interpretation, prevalent in the 1970s but virtually absent from discussions in the 1980s, is what O'Riordan calls 'centralised authoritarianism'. This generally follows the logic of Garrett Hardin's 'tragedy of the commons' (1968) which suggested that resources held in common would be overused. This metaphor led to the argument that centralised global political structures would be needed to force changes in behaviour to reach sustainability.[22] In some versions, this involved the adoption of what were called 'lifeboat ethics' (Hardin, 1974). The idea was that the scarcity outlined by Meadows *et al.* meant that rich countries would have to practise triage on a global scale – to 'pull up the ladder behind it'. This argument has however been rejected by Greens, with a few exceptions.

The third position is similar to the above in that it suggests authoritarianism may be required, but rejects the idea that this can be on a global scale. The vision here is for small-scale, tightly knit communities run on hierarchical, conservative lines with self-sufficiency in their use of resources (e.g. Heilbroner, 1974; The Ecologist, 1972). It shares with the above position the idea that it is freedom and egoism which have caused the environmental crisis, and these tendencies need to be curbed to produce sustainable societies. In some versions, these communes would be inward-looking and explicitly xenophobic (e.g. Hunt, n.d.).

The final position which O'Riordan outlines is termed by him the 'anarchist solution'. This has become the position adopted by Greens as the best interpretation of the implications of limits to growth. The term 'anarchist' is used in this typology loosely. It means that Greens envisage global networks of self-reliant communities.[23] This position would, for example, be associated with people like E.F. Schumacher (1976), as well as bioregionalists such as Kirkpatrick Sale (1980).[24] It shares the focus on small-scale communities with the previous position, but has two crucial differences. Firstly, relations within communities would be libertarian, egalitarian and participatory. This reflects a very different set of assumptions about the origins of the environmental crisis; rather than being about the 'tragedy of the commons' (which naturalises human greed), it is seen to be about the emergence of hierarchical social relations (Bookchin, 1982), and the channelling of human energies into productivism and consumerism. Participatory societies should provide means for human fulfilment which do not depend on high levels of material consumption. Secondly, these communities, while self-reliant, are seen to be internationalist in orientation. They are not cut off from other communities, but in many ways conceived of as embedded in networks of relations of obligations, cultural exchanges, and so on.[25]

Greens also often object to the state for anarchist reasons. For example, Spretnak and Capra (1985, p. 177) suggest that the features identified by Weber as central to statehood are often the problem from a Green point of view. Bookchin (1980; 1982) gives similar arguments, suggesting that the state is the ultimate hierarchical institution which consolidates all other hierarchical institutions. Carter (1993) suggests that the state is part of the dynamic of modern society which has caused the present environmental crisis, as outlined earlier in this chapter.[26] Thus the state is not only unnecessary from a Green point of view, it is positively undesirable.[27]

Whether or not we term the result 'anarchist' (which often ends up as a simple terminological dispute about what we mean by the 'state'), the dominant political prescription within Green politics is for a great deal of decentralisation of political power to communities much smaller in scale than nation-states, and for those communities to be embedded in networks of communication and obligation across the globe (that is, for them to be non-sovereign). Such political decentralisation would be accompanied by a scaling-down of economic activity, with a much greater proportion of needs being met locally, and global economic exchanges being correspondingly curtailed and left to those which either are necessary for global cultural connections to be maintained (so that communities do not become parochial) or for things which can only be produced outside a particular region and where living without them would significantly impair quality of life. Greens are often unclear, however, on whether such economic decentralisation also entails significant change in production relations (although by their practices they reveal a clear interest in communal, cooperative work relations). Greens also suggest that revitalising small-scale communities creates the possibility of moving away from social ontologies based on domination, for 'cutting off the King's head'. One of the reasons Dryzek gives in support of his arguments in favour of decentralisation is that (1987, p. 219) such small communities are more likely to develop a social ontology which undermines purely instrumental ways of dealing with the rest of nature (because their dependence on local resources is so immediately obvious to them).[28] This move away from modes of domination would also be possible with regard to relations between people, challenging entrenched patriarchal modes.

One specific form of such an argument for decentralised, egalitarian, non-sovereign communities can again be seen in writers like the Ecologist, in their focus on 'reclaiming the commons' (Ecologist, 1993).[29] The argument is essentially that common spaces are sites of the most sustainable practices currently operating. They are under threat from development which continuously tries to enclose them in order to turn them into commodities. Therefore a central part of Green politics is resistance to this enclosure. But it is also a (re)constructive project – creating commons where they do not exist.

Commons regimes are difficult to define, as The Ecologist suggests. In fact they suggest that precise definitions are impossible, as the variety of commons around the world defy clear description in language. The first point of definition is a negative one, however. The commons is not the commons as referred to by Garrett Hardin (1968). His 'tragedy of the commons', where the archetypical English medieval common gets overgrazed as each herder tries to maximise the number of sheep grazed on it, is in practice not a commons, but an 'open access' resource (The Ecologist, 1993, p. 13).

Commons, therefore, are not anarchic in the sense of having no rules governing them. They are spaces whose use is closely governed, often by informally defined rules, by the communities which depend on them.

They depend for their successful operation on a rough equality between the members of the community, as sharp imbalances in power would make some able to ignore the rules of the community. They also depend on particular social and cultural norms prevailing; for example, the priority of common safety over accumulation, or distinctions between members and non-members (although not necessarily in a hostile sense, or one which is rigid and unchanging over time) (ibid., p. 9).

Commons are therefore clearly different from private property systems. However, commons are also not 'public' spaces in the modern sense. Public connotes open access under control by the state, while commons are often not open to all, and the rules governing them do not depend on the hierarchy and formality of state institutions. A further difference from 'modern' institutions is that they are typically organised for the production of use values rather than exchange values, i.e. they are not geared to commodity production. This makes them not susceptible to the pressures for accumulation or growth inherent in capitalist market systems.

Commons are therefore held to produce sustainable practices for a number of reasons. First, the rough equality in income and power means that none can usurp or dominate the system. 'Woods and streams feeding local irrigation systems remained intact because anyone degrading them had to brave the wrath of neighbours deprived of their livelihood, and no one was powerful enough to do so' (The Ecologist, 1993, p. 5). Second, the local scale at which they work means that the patterns of mutual dependence make cooperation easier to achieve.[30] Third, this also means that the culture of recognising one's dependence on others and therefore having obligations, is easily entrenched. Finally, commons make practices based on accumulation difficult to adopt, usufruct being more likely.

One of the great strengths of The Ecologist's work is the way in which the argument is richly illustrated. I will give just a few examples here. At a general level, they highlight how many people throughout the world are dependent on commons, despite the globalisation of capitalism. For example, 90 per cent of the world's fishers depend on small inshore marine commons, catching over half of all the fish eaten (The Ecologist, 1993, p. 7; Ostrom, 1990, p. 27).

In the Philippines, Java and Laos, irrigations systems are run by villages communally, with water rights decided at the village level. Even in the North, they suggest, communities still exist which manage resources communally – for example, lobster harvesters in Maine (The Ecologist, 1993, p. 7). In parts of India, villages based on

Gandhian principles, known as *gramdam* villages, enable sustainable practices to flourish. In these villages, all land within the village boundary is controlled by the *gram sabha*, composed of all the adults in the village (ibid., p. 190). They quote Agarwal and Narain on how one such village, Seed near Udaipur, operates:

> The common land has been divided into two categories – one category consists of lands on which both grazing and leaf collection is banned and the second category consists of lands on which graz-ing is permitted but leaf collection or harming trees is banned. The first category of land is lush green and full of grass which villagers can cut only once a year. ... Even during the unprecedented drought of 1987, Seed was able to harvest 80 bullock cartloads of grass from this parch. The grass was distributed equitably amongst all households.
>
> (Ibid., pp. 190–1; see also Agarwal and Narain, 1989, p. 23)

The idea of the commons is clearly very consistent with the arguments from Green political theory about the necessity of decentralisation of power, and grassroots democracy. It shows how small-scale democratic communities are the most likely to produce sustainable practices within the limits set by a finite planet.

Simultaneously, however, the focus on the commons helps to start thinking about the second question outlined at the beginning of this section. It suggests, along with other arguments presented here, that the sites where global environmental politics can be pursued are less in the halls of the UN in Geneva or New York than at the sites where commons regimes, and more broadly small-scale local communities, are being destroyed by development or being consciously brought into being. Such a politics is therefore necessarily one of resistance, where resistance is understood as simultaneously reconstructive (not simply preventing 'bad' things happening, but active attempts to produce a 'better' future). I will revisit and develop this theme in Chapter 7. Now I turn to some empirical cases where I try to illustrate the argument developed so far. The next three chapters hope to show firstly that the global power structures outlined in this chapter systematically produce global environmental change; secondly, that the ways in which partic-ular political institutions are embedded in broader structures compli-cates their efforts to respond to such change; and thirdly, that the sites of global environmental politics should be understood as sites of resistance.

4
Space, Domination, Development: Sea Defences and the Structuring of Environmental Decision-Making

> I fear that to many members of this society the subject of Coast Erosion will appear one of those which, though important enough to make it a matter of satisfaction that a paper should be read on it, yet it is hardly interesting enough for them to take the trouble to come and hear the paper.
>
> (Bourdillon, 1886, p. 1)

What this chapter tries to show is that particular decisions concerning global environmental change are deeply conditioned by the power structures I discussed in the previous chapter. I develop these arguments through an analysis of the history and contemporary politics of sea defences in Eastbourne (a medium-sized town on the UK's south coast) in particular, and the UK more generally. I discuss below the particularities of the decisions surrounding the replacement of Eastbourne's sea defences in 1994–97. I try to show how the particular decisions were structured by a variety of political, economic and discursive constraints on both Eastbourne's local government bodies and on the producers of the timber in Guyana. I will give a summary of those arguments later in this chapter. Before that, however, I want to develop more general points concerning the way that sea defences are themselves embedded in the specifically modern twin projects of the human domination of nature and of some humans by others, and in the dynamics of capital accumulation, or 'economic development'.

Projects of domination

> If the government can't protect its own coastline, we need a change of government.
>
> (Voxpop interview on *Channel 4 News*, 28 August 1996)

The image of sea defences is perhaps one of the clearest images of the human domination of nature. The physical control of one of nature's most powerful forces, water, exemplifies the project of mastery in symbolic terms as much as anything. In the Anglo-Saxon world, the story of Canute, King of England in 1016–35, who (at least in popular mythology) tried to force the sea to recede by virtue of his monarchical authority and literal command, has also served as a metaphor for the ways that defences against the sea have symbolically been bound up with twin projects of domination: of nature by humans, and of space by increasingly territorially defined political units, culminating in modern nation-states. These discourses are also closely interconnected with processes of economic development and capital accumulation. Often, as I will show regarding Eastbourne, the direct impetus for building the first sea defences was protection of property. But the need to protect property in this way was produced by development as it pushed people closer to the shore. In Eastbourne's case this was because of the development of an urban elite in Victorian Britain which made possible the thriving south-coast resorts, of which Eastbourne was one among many, as those elites escaped London in the summer for fresh sea air. However, the rhetoric of sea defences in Eastbourne remains bound up with military metaphors where the sea is equated with human enemies. And the modern concept of property which made possible the notion that there was a thing to be defended (property being the thing itself, rather than what one had in the thing), is directly related to territorial and sovereign notions of the nation-state; both modern notions of private property and state sovereignty coevolved as mutually reinforcing concepts based on absolute, exclusionary rights (Burch, 1994).

The symbolic aspects of sea defences directly concern the twin projects of the domination of nature and ongoing nation-building. In Eastbourne, as elsewhere, sea defences show how these two projects are intertwined with each other. The starkest aspect of this is in the use of military metaphors. The very term 'sea defences' is of course fundamentally a military construction, connoting defences against an enemy with agency.

Historical accounts of sea defences in Eastbourne are explicit in this regard. For example:

> The Second World War saw the erection of the most long lasting and systematic set of defences against invasion by the enemy which have ever been seen. Unfortunately, that other enemy, the sea, had to be left to do its worst.
>
> (East Sussex County Planning Department, 1977, p. 3)

Wright also invokes such metaphors. 'The sea was regarded with some distrust at the beginning of the second half of the last century', he writes (1902, p. 163). The 'encroachments' of the sea in this period involved many instances where houses were 'demolished by the inroads of the invader' (ibid.). The builder of much of the late nine-teenth century defences, a Mr Berry, is regarded in heroic terms 'as a pioneer of sea-wall defence' (ibid., p. 164). In more recent times, public representation of the sea-defences question continues to be couched in military terms, with headlines such as 'Battle against the Sea' (*Eastbourne Gazette*, 28 November 1984). Potent symbols, playing on the historical role of the south coast in East Sussex in the last invasion (by a human army) of England, are sometimes used. Consider this line in the middle of a general discussion of the replacement of sea defences east of Eastbourne: 'When William the Conqueror set foot on English soil in 1066, Pevensey was on a peninsula and the sea extended as far inland as modern Hailsham' (*Bexhill Observer*, 12 April 1996).

When the contemporary system in the UK for organising sea defences was instituted, in the Coast Protection Act 1949, one of the intentions was to enable the reclamation of land for agricultural purposes. It was connected closely to the strategy of food self-sufficiency adopted during and after the Second World War, to limit the UK's vulnerability (as an island) during wartime and thus contribute directly to the UK's capacity to wage war. The possibility of adopting a 'managed realignment' policy (see below) has been premised on the 'lessened need for agricultural self-sufficiency', as such understandings of the UK's strategic needs have waned (Agriculture Committee, 1998a, p. xxxviii).

Metaphors of military battle and nation-building are prevalent in human attempts to deal with some of the problems caused by water. The classic case is the Netherlands. Simon Schama (1987) shows how the defence against the sea was intimately connected with the construction of a nation. During the late medieval period, the Netherlands experienced some of its biggest floods. And the processes of national resistance against Spain and against the sea were closely linked in the popular imagination during the struggle for indepen-dence in the late sixteenth century (ibid., p. 37). After independence, nation-building was associated with claiming new land from the sea. Land reclamation was always, according to Schama, associated by the Dutch with the construction of a national identity. Dutch identity was formed by a 'trial of faith by adversity', in which saving land from water was one of the central elements. Their faith in God was to be tested by trials, in particular imposed by the sea, but on the other hand, their capacity to reclaim land from the sea was given scriptural significance, indicating a special favour with God. Schama cites Andries Vierlingh, a sixteenth-century hydraulic engineer: 'The making of new land belongs to God alone ... for he gives to some people the wit and the strength to do it' (Schama, 1987, p. 44).[1] Thus a Christian nation could be legitimised and forged. In modern times, such a linkage is still present. '"In many ways, our dikes are part of our identity", said Hans Scholten of the Dutch water authorities. "They're part of being a Dutchman – not just our salvation, but our pride"' (*Guardian*, 2 February 1995, p. 9).

John McPhee's (1989) account of the control of the Mississippi also tells a dramatic story of how understandings of human dealings with water are directly understood in military terms. His story focuses on the controls upstream of New Orleans designed to prevent the main course of the river from moving West, as it would if uncontrolled by humans. The story is an epic tale of human domination of nature (one of the people he talks to calls it the 'third greatest arrogance' of humans in their dealings with nature [1989, p. 99]), in which nature is directly understood as an enemy. It is also one where human dealings with nature are directly militarised; responsibility for controlling the course of the Mississippi along its length rests with the US Army Corps of Engineers.[2]

'Nature, in this place, had become an enemy of the state', he writes (ibid., p. 95). Much like East Sussex County Council Planning Department (1977, see more below), the Army lockmaster in charge of the river control schemes, understands the river as an enemy:

> This nation has a large and powerful adversary. Our opponent could cause the United States to lose nearly all her seaborne commerce ... We are fighting Mother Nature ... It's a battle we have to fight day by day, year by year; the health of our economy depends on it.
>
> (Rabalais, quoted in McPhee, 1989, p. 95)

If the river is the enemy, then the control structure, known as Old River Control, is 'a good soldier', in the words of an Army promotion film (ibid., p. 100). And floods, which threaten to destroy the control structures, must be the battles. 'In 1983 came the third-greatest flood of the twentieth century – a narrow but decisive victory for the Corps' (ibid., p. 141).

The military metaphors central to understandings of sea defences are best understood in terms of nation-building. The construction of a nation under modernity has been a physical manifestation of national sovereign power, as concrete expressions of a state's capacity to defend 'its' territory. The territoriality involved here is specifically modern. The emergence of mapmaking as an activity also helped generate these practices, by solidifying the sense of where a nation's shores began and ended.

It is often noted in much contemporary IR theory that boundaries are central to state-building. The boundary is the key site at which identity is constituted through difference. In Devetak's words, 'the political begins with the practices that inscribe and administer the boundary which establishes the space for political institutions' (1995, p. 30). But nature, and perhaps particularly the sea, are important correlates here, usually overlooked. Margin (another commonly used term for a boundary, celebrated in poststructural IR theory), as Stilgoe points out (1994, p. 9) derives from 'marge', an antiquated word for the coast or shore. And the sea is directly analogous to the anarchy which is the external correlate of sovereignty and order. The sea is the nature which cannot be domesticated, tamed, farmed. So while 'boundaries function to divide an interior, singular, sovereign space, from an exterior, pluralistic, anarchical space' (Devetak, 1995, p. 30, citing Walker, 1990), the boundary of the land and the sea is a double boundary – between 'inside and outside' in political terms (Walker, 1993), and between a human-dominated 'nature' and that 'nature' which humans cannot dominate.

So nation-building is simultaneously a part of the human domination of nature, of the imposition of human physical control over land and water. An opposition between humans and nature is presupposed in those involved in building such structures. In the Mississippi, the General in charge of Old River Control states that:

> had man not settled in southern Louisiana, what would it be today? Under nature's scenario, what would it be like? ... If only nature were here, people – except for some hunters and fishermen – couldn't exist here.
>
> (quoted in McPhee, 1989, p. 145)

Nature must therefore be controlled in order to make a flourishing of human life possible.

Sea defences (as is most obvious in the Netherlands) developed concurrently with the development of the modern science of Bacon and Descartes. In many ways, they provide a clear physical manifestation of the ways in which modern scientific knowledge was connected to discourses of the domination of nature – connections which Bacon made explicit – and which as Merchant (1980) and others suggest, was specifically gendered in the way that women became associated with nature.

In this context, the portrayal of the sea as active in many of the accounts given above is a surprisingly pre-modern notion of nature. However, it provides the background against which the sea becomes something to struggle with, and ultimately to dominate. In accounts of sea defences, however, the sea, while it may in certain contexts have been pacified, it has never become passive. This has produced a particular version of the domination of nature which involves a certain subtlety and art. Schama cites Vierlingh, a Dutch writer of the mid-sixteenth century, as developing a 'humanist philosophy of hydraulics':

> Reason is preferred to force. If the waters are met by mere barriers, they will repay that *fortse* with interest. Instead, the 'persuasion' of streamlining and channel cutting can in effect civilize the waters.
>
> (Schama, 1987, pp. 42–3)

Concerning Eastbourne, Wright quotes an article in the *Eastbourne Gazette* which shows that Berry, his 'pioneer of sea-wall defence', had a similar sensibility. He built a wall set at an angle from the vertical. 'Instead, therefore, of the sea striking a dead perpendicular wall, it expended its energy in running up the slope. His contention was that the constant hammering of the sea against a perpendicular wall must sooner or later shake it to pieces, and the only safe plan was to *humour it* in the way he carried it out' (Wright, 1902, p. 164, emphasis added).

This particular history of the domination of nature in relation to sea defences has a curious ongoing footnote. From the 1970s onwards, in part in response to the emergence of environmentalism, council planners and others appropriated holistic notions derived from environmentalism to discuss sea-defence planning. In 1977, East Sussex County Planning Department wrote that 'The lesson to remember is that nature only tolerates a certain amount of opposition, so it is best not to interfere too much or too long with the natural process, but instead to make use of it. If this were done, we should not always think of repelling waves but we would should think instead of using their energy to generate power' (1977, p. 130). During the 1990s, UK public policy has not taken the latter piece of advice, but has begun to treat sea defences in a manner involving holistic rhetoric.

Such rhetoric involves a recognition that sea-defence structures often cause as many problems as they solve. One such set of problems concerns the unintended consequences of projects further down the coast. The draft Shoreline Management Plan for Beachy Head to South Foreland (the stretch of coast including Eastbourne) states that 'traditionally, coast and flood protection works have been implemented in a piecemeal fashion. ... The possibility that measures taken on one section of the coast might affect adjoining areas was either not recognised, or not considered in sufficient depth' (BMT Limited, n.d.). There is a particular piece of historical forgetting here. Bourdillon, in 1886, was acutely aware of such effects along the coastline, referring to how a new breakwater at Newhaven (substantially west of Eastbourne) had the effect of moving shingle further along the coast and eroding land near Eastbourne further (Bourdillon, 1886, p. 9). But such a 'new' holistic awareness is substantially changing UK public policy on sea defences.

'Managed retreat', 'managed realignment' and occasionally 'strategic withdrawal' have become watchwords in this discourse, with military metaphors remaining, but the capacity of humans to win 'wars' against the sea having been conceded. The Shoreline Management Plan already referred to discusses 'Setting Strategic Coastal Defence Options', and gives three such options: to maintain the *status quo*; to 'develop new defences seaward of an existing line'; or to 'pull back from an established line and construct new defences inland of existing defences' (BMT Limited, n.d.). In many ways, however, this is perhaps a continuation of the way that Vieringhe or Berry discussed dealing with the sea; still a discourse of domination, but a subtler one than simply using brute force.[3]

The House of Commons Agriculture Committee held hearings and produced a report on Flood and Coastal Defence in 1998. The first paragraph of their conclusions and recommendations is worth quoting at length, recognising as it does the intertwining of militarism with sea defences, while also reflecting how partially the discourse of managed retreat has escaped this set of understandings:

Our nation's history is one of continual intervention in coastal and riverine processes, punctuated by occasional awesome reminders of the power of the sea. It is not surprising that our lexicon for describing the relationship between the land and the sea is dominated by militaristic terminology: we speak of flood and coastal 'defence', or 'reclaiming' or 'winning' land from the sea – even of the sea 'invading' the land. Hard-engineered defences remain essential to protect many vital national assets, especially in urban areas. But, overall, we believe it is time to declare an end to the centuries-old war with the sea and to seek a peaceful accommodation with our former enemy. It is far better to anticipate and plan a policy of managed realignment than to suffer the consequences of a deluded belief that we can maintain indefinitely an unbreachable Maginot Line of towering sea walls and flood defences.

(Agriculture Committee, 1998a, pp. xxxvii–xxxviii)

Such a move is therefore an attempt to move away from understanding the relationship with the sea in military terms. In such a shift, nature becomes 'our' friend rather than enemy – the focus is on 'enhancing natural coastal defence' (Agriculture Committee, 1998a, p. vi) through promoting salt marshes and the like, which limit the extend of flooding. The shift to managed realignment also perhaps reflects other discursive shifts – from 'hard' to 'soft' engineered defences, promoting 'flexibility' in the coastline (ibid., p. xxxii), both being resonant with post-Fordist managerial discourses. Such a shift also requires government to attempt to 'overcome the widespread public culture which is intolerant of the acceptance of naturally occurring and unavoidable risk associated with flooding' (ibid., p. vii), reflecting shifts in understanding of national space as homogeneous to one where people are being mobilised to be tolerant of difference. Now, clearly it would be overambitious to make causal claims concerning the importance of such phrases. But it does support the claim often made by Andrew Ross (1991) and others that scientific projects are always simultaneously as dependent on social context as they are on scientifically rational enquiry. So just because the militarism associated with 'hard' sea defences may be waning, its replacements will have as much ideology embedded in them. These replacement discourses ('soft', 'flexible', 'tolerant', 'holistic' sea defences) are still about an ongoing project of nation-building – but one which is less consistent with the territorially preoccupied state than with the post-Fordist competition state appropriate to the age of global capitalism, with a dose of multiculturalism thrown in.

Nation-building has produced a progressive centralisation of power in national-level political institutions. The emergence of increasingly centralised state systems over the last several centuries has left local political units with very little control over their own affairs, and with great incentives to operate in ways in which central governments dictate. In Eastbourne (see below), well over half of the money for the sea defences was to be provided by central government, through the Ministry of Agriculture, Fisheries and Food (MAFF). The MAFF also provided the criteria according to which the scheme should be judged, and provided other constraints, such as the fact that they only fund capital works, not renovation of existing defences (providing an incentive to replace whole sea-defence arrangements rather than maintain existing ones). That state centralisation, is often associated, for example by many historical sociologists (e.g. Tilly, 1990; Mann, 1986) with the emergence of a competitive states system. This has had the effect of creating incentives for states to develop continuously more efficient means of extracting resources from a territory and a population in order to be able to engage in warfare. Writers such as Braudel have highlighted the ecological consequences of this state-building (see, for instance, Helleiner, 1996), but here the major point is how centralisation constrains the capacities of political units below the nation-state level to respond to environmental problems themselves.

In the UK sea defences were effectively nationalised (in the sense of the emergence of a nationally standardised system regulated by central government rather than simply organised on an *ad hoc* basis by local government) in 1949 by the Coast Protection Act of that year (Trafford and Braybrooks, 1983). What is missed in Trafford and Braybrooks' account, however, is that this coincided with a general centralisation of British government as the postwar Labour government nationalised various industries. This was as much a matter of taking industries from local political units into central control as it was of moving industries from the private to the public sector, and also involved the increasing standardisation of many aspects of public life (including education and health) across the country. It is likely then that the Coast Protection Act was not purely a function of recognition of the 'need' for nationally coordinated sea defences, but was also produced by more general discursive shifts. The knock-on effects of this centralisation, however, were one of the features of the debate in Eastbourne in the early 1990s.

Development and the origins of sea defences

Historically, the construction of sea defences appears to be bound up primarily with local economic forces. There are no clear sources giving the dates of the earliest sea defences in Eastbourne, but historians of both sea defences and of Eastbourne suggest they began in the early nineteenth century. (In general, while use of some sorts of sea defences, such as land drainage, goes back further, use of groynes as sea defences in Europe began in the seventeenth century – see Brampton and Motyka, 1983).[4] Tyhurst infers this from comparison with Brighton, then a much bigger town further west along the coast. In 1705, after two very damaging storms, Brighton first considered building sea defences (Tyhurst, 1972, p. 2). These defences were later built in 1723 (East Sussex County Planning Department, 1977, p. 2). Chambers, writing in 1910, writes that 'The construction of groynes [in Eastbourne] appears to have begun some time in the early part of the 19th Century' (1910, p. 212). Chambers also says that the first sea-wall along part of Eastbourne's sea-front was built in 1847–48, and that before that there were only a few groynes as defences against the sea. Wright, writing at about the same time as Chambers, also says that by 1852 there was a small sea-wall and a number of groynes (1902, p. 164). These groynes first appeared in maps of the town in the 1875 edition of the Ordnance Survey map of the area (Tyhurst, 1972, p. 2).

These defences were then rapidly expanded in the later nineteenth century. By 1900, there were 63 groynes along the promenade at Eastbourne, and a few more further east. A few more were built in the early twentieth century. There was also a greatly expanded sea-wall, made possible by an Act of Parliament of 1879 which gave the local board the authority to build such a wall (Chambers, 1910, p. 215; also Enser, 1976, p. 22).

The early constructions were to protect specific properties from damage by the sea. Wright recounts a conversation with a 'Miss Ingledew' concerning the construction of the wall in 1847–48. This was specifically to protect the houses opposite the Marine Parade (Wright, 1902, p. 92). Chambers gives detailed accounts of the storms of 1876 and 1877, the latter of which 'wasted away part of the Pier, flooded many houses at the seaside, and caused a great devastation of property' (1910, p. 215). These storms gave the impetus for expanding the sea wall in the following years.[5]

But a purely functional interpretation of the construction of sea defences in Eastbourne would be inadequate. Firstly, the function that sea defences performed in protecting property was interconnected with the meanings that property had come to have by the nineteenth century, a meaning that had evolved noticeably since at least the seventeenth century. Property had ceased to be a particular right that a person had in a thing (for example, a right to use particular resources from a particular plot of land) and had become the thing in itself (exclusive rights to the land in its entirety of uses) (Macpherson, 1975; Reeve, 1986). This is what was involved in the commodification of land associated with the emergence of capitalism. Thus, during this period (around the seventeenth century), 'states acted to establish bourgeois property rights – to reduce multiple claims on the same property, to enforce contracts, and to strengthen the principal owner's capacity to determine the property's use' (Tilly, 1990, p. 100). One of the consequences of this shift was that people who owned land could only meet certain needs through the land they owned, whereas they might previously have been able to meet them through their diverse properties in a wide area of land (The Ecologist, 1993, pp. 22–4). It meant they had a far greater dependence on particular plots of land and thus a stake in protecting them.

Simultaneously, the nineteenth century saw a great expansion of both population in general in Western Europe, stimulated by the Industrial Revolution, among other things, but also of a bourgeois class with significant leisure time, living in cities which were increasingly unpleasant environments. Thus many coastal towns in the UK became resorts for that class to retire to for substantial stretches of the summer, both for recreational and therapeutic purposes. Thus the values of properties along the coast rose, and along with them the stakes involved in protecting them from the sea. So the functional aspects of sea defences are closely interconnected with a historically specific process of economic development or capital accumulation associated with Eastbourne's economy in the nineteenth century and with chang-ing notions of property. In Brampton and Motyka's terms, discussing the rise in groyne construction in the mid–late nineteenth century, this coincided 'with the development of coastal resorts and seaside housing in general' (1983, p. 151).

As in Eastbourne, a direct link to processes of economic development is also present in the tales of both Schama and McPhee. In the Netherlands, land reclamation enabled Amsterdam capitalists to make large amounts of money in land-speculation on the reclamation pro-jects, and thus to stimulate the projects (for example by providing money) in order to make money for themselves (Schama, 1987, pp. 38–9). And on the Mississippi, much of the immediate impetus to continue preventing the river course from changing is economic. The stretch of the Mississippi between Baton Rouge and New Orleans used to be known as 'the American Ruhr' because of the amount of heavy industry located there – 'B.F. Goodrich, E.I. du Pont, Union Carbide, Reynolds Metal, Shell, Mobil, Texaco, Exxon, Monsanto, Uniroyal, Georgia-Pacific, Hydrocarbon Industries, Vulcan Materials, Nalco Chemical, Freeport Chemical, Dow Chemical, Allied Chemical, Staufer

Chemical, Hooker Chemicals, Rubicon Chemicals, American Petrofina' (McPhee, 1989, p. 94). The industries were there because of the river and the services it provided – water, transport, and so on. Those industries would be devastated without the steady flow of the Mississippi. In McPhee's words, for economic reasons, 'for nature to take it course was simply unthinkable. The Sixth World War would do less damage to southern Louisiana' (ibid., p. 95). Army geologists thought the major purpose of the river control was economic (ibid., p. 180).

Structuring environmental decision-making

It is in this sort of context that we should understand political decision-making concerning global environmental change. In the mid-1990s in Eastbourne, there was a controversy over the way the local Council planned to replace its sea defences.

The replacement of Eastbourne's sea defences

In April 1995, a small news story in the UK media reported that Eastbourne Borough Council was planning to import greenheart, a tropical hardwood, from Guyana, to rebuild its sea defences. These defences needed to be improved because of worries about global climate change, especially its associated sea-level rise and the possible increased frequency of stormy behaviour. The story focused on the inevitable irony that while defending against a global environmental threat, the Council was simultaneously contributing to causing that threat, through the related problem of deforestation (*Channel 4 News*, 11 April 1995).

The immediate origins of the Council's decision were in the heavy storms which occurred in Southern England (and elsewhere) in the winter of 1989–90. These storms caused much damage to the groynes along the seafront at Eastbourne.[6] There was also further storm damage to the groynes later, especially during the winter of 1992–93, while the plans for renovating or replacing the sea defences were being drawn up. These later storms helped the Council to emphasise the need for completely new structures to be built, rather than old ones repaired. In 1990, the Council appointed Posford Duvivier as its consulting engineers to assess what possibilities there were in replacing the system of defences. Posford Duvivier produced a number of reports for the Council's committees over the next three years, advising the Council on the best course of action. These narrowed down the most desirable options from an original list of six to a recommended three (timber, rock or concrete groynes). Of these, the Council's Strategic and

Economic Development sub-committee of its Policy and Resources Committee rejected one immediately (an option of building concrete groynes), and asked Posford Duvivier to work more on the remaining two options. By early 1993, Posford Duvivier produced another report recommending the construction of timber groynes. They were also by now recommending the use of greenheart as the most appropriate timber from an engineering point of view. They were then asked to produce a scale model of the new structures and test it, to improve the design, and reduce costs. This they did during 1993 and reported back to the Council on the results of the modelling in early 1994. After the modelling results were shown to the Council, the plans were sent to the Ministry of Agriculture, Fisheries and Food (MAFF) for approval.[7]

At this point, the Council opened the plans to formal public consultation. Quickly, the local Friends of the Earth (FoE) group and some others got involved, to protest about the use of tropical hardwoods in the project. Michael Le Page of the local FoE group, who organised much of the campaign, managed also to involve Simon Counsell, then Forest Campaigner for FoE UK in London. Their campaign (along with the time taken to get MAFF approval) clearly slowed down the project; the intended start of construction was pushed back from late 1994 to mid-1996.

There were a number of points of contention in the conflict between FoE and the Council/Posford Duvivier. The main one concerned the sustainability of the production of greenheart timber. The main two points here in the public debate were the rates of regrowth of greenheart trees, and the question of how 'sustainability' should be assessed. The campaigners argued that only timber produced according to Forestry Stewardship Council (FSC) standards should be used, as this was the only independently verifiable certification of sustainability. The Council and Posford Duvivier argued that the company they were intending to buy the timber from, Demerara Timber Ltd, (DTL) had their own Green Charter, independently verified by SGS Forestry (in the sense that SGS said that DTL did actually do what it said in their Charter).[8] In addition, SGS Forestry were soon to be one of the companies approved by FSC to accredit timber operations as FSC accredited. Nobody at this point explicitly claimed, however, that DTL's Green Charter was the same as FSC criteria.

The second main conflict was over the availability and suitability of oak. Posford Duvivier claimed early on that 'although it had originally been thought that English oak could be used, there was currently insufficient available at economic cost for a scheme as large as that now proposed for Eastbourne' (Eastbourne Borough Council, 1993). Local timber producers, supported by the local FoE group, and by East Sussex County Council's Environmental Planners (the County Council having responsibility for forests in the area, including promoting the economic use of those forests) contested this claim, and argued that they were never given the opportunity to tender, since the consulting engineers wrote the specifications in such a way that greenheart was advantaged from the outset. They also claimed that oak and greenheart were never treated on a comparable basis; for example, claims by Posford Duvivier that oak was not strong enough for long-lasting construction were based on experience in using nine-inch square sections of timber, whereas all the projections for using greenheart were on the basis of using twelve-inch sections, and that previously relatively cheap-quality oak had been used, whereas higher-grade oak timber could be supplied. Furthermore, they claimed that the existing oak groynes needed replacing not because the wood was inadequate: 'the failure of the present groynes is principally due to the corrosion of the metal ties which hold the timbers together. They have rusted badly. The timber itself has not failed…' (Ray Russell, quoted in *Sussex Express*, 6 January 1995).

The outcome was that Posford Duvivier and the Council's engineers got what had been their 'preferred option' since early 1993 approved and used. Construction, using greenheart, commenced in mid–late 1996. Despite the protests, greenheart has been used throughout the project, even without FSC certification. The engineers involved were clearly keen to legitimise their project in the face of environmental objections, but were also clearly more keen to use greenheart in any case. In their eyes, the superiority of greenheart over oak from an engineering point of view was the trump card. In Council meetings in March 1993, January 1994 and June 1994, the engineers made significant efforts (even before FoE began its campaign) to assure councillors that the timber would come from sustainable sources, and spent effort and money trying to find sources that could plausibly be shown to be 'sustainable'. In June 1995, after extensive campaigning by FoE and others, the Council agreed only to use timber which was FSC approved. However, Posford Duvivier and the Council's engineers then found that there were at that time no sources of FSC-accredited greenheart (or, for that matter, FSC-accredited oak), and none was likely to be available by the time that construction was then scheduled to start (early 1996). DTL had agreed to apply for FSC accreditation, but SGS forestry was not yet accredited by the FSC to accredit timber as FSC approved. Then DTL was taken over by a Singapore-based company and this further delayed the process of getting their wood FSC approved. So the engineers went back to the Council, who reversed their commitment to use only FSC-approved timber, in meetings in November 1995 and January 1996, arguing by now, however, that DTL's Green Charter was adequate.[9]

Structuring environmental decision-making

The ironic nature of this situation was lost on no one. While trying to respond to 'threats' imposed by the environment, which were increasingly being connected in people's consciousness to large-scale environmental change such as global warming, the Council was simultaneously helping to aggravate those very environmental changes in the first place. Much of the force of campaigns by Friends of the Earth (FoE) and others locally came from the recognition in people's minds and in public discourse of this ironic dilemma.

At the level of political decisions, this dilemma and dynamic runs throughout global environmental politics. This dilemma is, however, itself a product of the complexity of modern, globally organised political, social and economic systems. The capacity of a Council to be enmeshed in global networks which make its decisions and their ramifications so convoluted shows many of the features of such a globally organised complex system.

This complexity could be understood in the terms outlined by Ulrich Beck in his analysis of how risk works in modernity. For Beck, risk is a pervasive feature of modern, large-scale technological systems. But what risk in these systems does is obscure lines of causality, which in turn make claims about responsibility for action difficult to sustain and easy to avoid. For Beck, this can be described as 'organised irresponsibility'. In modern societies, risk is defined legalistically in order to try to assign culpability for causing damage. But under conditions where the hazards and risks are widespread, and caused by multifarious agents and forces:

> the interpretation of the principle of causation in individual terms, which is the legal foundation for hazard aversion, protects the perpetrators it is supposed to bring to book...an ostensibly protective judicial system, with all its laws and bureaucratic pretensions, almost perfectly transforms collective guilt into general acquittal.
>
> (Beck, 1995, p. 2)

In other words, in Goldblatt's terms, organised irresponsibility 'denotes a concatenation of cultural and institutional mechanisms by which

political and economic elites effectively mask the origins and consequences of the catastrophic risk and dangers of late industrialization'
(Goldblatt, 1996, p. 166).

One of the consequences of this is that much of the conflict over
political decisions with respect to the environment will concern
notions of responsibility and agency. One aspect of this can be seen in
representations of structure and agency in this regard. While Channel
4's narrative was structural, prevalent stories in Eastbourne itself were
agency-oriented. From FoE, the dominant construction was that
Eastbourne Borough Council was destroying a rainforest; their leaflet
asking people to write to the Council protesting had the slogan 'Visit
the Rainforest...come to Eastbourne'. Indeed, Simon Counsell, when
questioned on this, responded neatly that 'if the Council's hands really
were tied, it's amazing to see the efforts they went to to tie their hands
up' (interview, 29 November 1996).[10] This is consistent with Beck's
argument that what risk under complexity does is effectively make it
too difficult to attribute blame or causality, allowing agents to evade
responsibility for their actions.

But the dominant public discourse in the local media was also
agency-oriented, concerning the Council's obligations both to protect Eastbourne from the sea, but also not to damage the global
environment. Engineers from the Council were perhaps more equivo-
cal in this regard; on the one hand, they ardently argued that they
had made the right decision and when asked about constraints producing their decision, denied there were any significant ones (apart
from the statutory obligation to provide sea defences), but on the
other hand, their arguments were clearly embedded in an engineering
discourse, such that the 'logic' of the engineering requirements for
the groynes predetermined a necessary outcome (involving the use
of greenheart). The structure determining their decision was what
the engineering required; their agency was that they were good
engineers.[11]

My concern here is to situate this particular, local political decision
within broader, globalised contexts. Such structural and discursive
contexts helped to produce this particular decision. Interpreting the
event in this way can therefore tell us much about the dynamics
of global environmental politics and the possibilities of understanding
such politics in terms which recognise the need for broad social
change, rather than ones which prescribe institutional and policy
change, as is currently hegemonic within academic debates surrounding global environmental politics.

Local economies

Eastbourne's economy, and therefore the local state's finances, are characterised and structured by a heavy reliance on tourism. The Council spent £2.8m in 1996–97 on tourism-related activities or 24.9 per cent of total spending, indicating the substantial importance which tourism has for the Council (Eastbourne Borough Council, 1996–97). Consideration of aesthetics and how the sea-defence structure would affect the local tourist industry were therefore important in influencing the Council's decision.

Aesthetics were clearly important in the rejection of a number of options originally put forward by Posford Duvivier. Three options (rock groynes, simply improving the sea wall [and abandoning groynes altogether], and concrete groynes) were rejected either directly, because of 'loss of beach for tourism' (in the case of the improved sea-wall option), or indirectly, because of 'substantial visual impact', which it is reasonable to assume is a euphemism for worries about loss of tourist revenue, not least since the seafront in the town centre is predominantly hotels (Eastbourne Borough Council, n.d, p. 3).[12] In the Council's formal deliberations, of the three options outlined as most favourable – timber, rock, or concrete groynes – the Council's Strategic and Economic Development Sub-committee rejected concrete as 'it was considered that option 4B, concrete bastion groynes, would detract from the ambiance of Eastbourne seafront to an unacceptable degree' (Eastbourne Borough Council, 1992, p. 365).

There was a certain amount of slippage here which helped the Council and its contractors narrow down the options from rock or timber, to timber only. In formal documents, only concrete is eliminated because of aesthetic or tourist-related reasons. Rock is later rejected for 'technical and financial reasons' (Eastbourne Borough Council, 1993, p. 743).[13] However, in other representations, both concrete and rock are lumped together as being rejected for this reason. This occurs in the Council's information pack, cited above, where there is also a photograph of rock groynes in Arun looking dingy and with a sign saying 'Caution, keep off rocks, no climbing, no diving, no swimming', and a caption reading 'Rock groynes in Eastbourne would be larger (than wood or than Arun's?) and subject to weed growth' (Eastbourne Borough Council, n.d, p. 3). It also occurred in the *Channel 4 News* item, where it was suggested that one of the reasons why the Council chose timber over stone or concrete was the fact that it was, in Channel 4's words, 'more attractive to tourists' (*Channel 4 News*, 11 April 1995).

The most plausible interpretation of the effect of this slippage is that it helped legitimise the use of the timber. The pervasive understanding of the reliance on tourism in Eastbourne would mean that the Council did not have to fully argue a case that a particular option would be problematic from an aesthetic/tourist point of view; rhetoric such as that in the information pack would suffice. The Council certainly argued that the project as a whole was, in the words of Janet Grist, chair of the Environment Committee, 'vital for the protection of Eastbourne and will also significantly improve leisure facilities and the environment along the seafront for residents and visitors alike' (quoted in the *Eastbourne Herald*, 12 August 1995, p. 7). Objectors also occasionally made similar points: 'Groynes are the better, and more aesthetically pleasing alternative to other engineering solutions, especially for a town which relies on tourists who come to visit its beach' (Russell, n.d, p. 1).[14] The tourism argument appeared to have been an implicit trump card in deciding the materials to be used, legitimising the decision to use wood in such a way that technical, financial or environmental arguments could not challenge its discursive force.

This discursive force is revealed in an editorial in the *Eastbourne Herald*. The editorial starts by discussing the replacement of the groynes, focusing on experts' agreement over global warming, Eastbourne's particular vulnerabilities, the controversy over the groyne timber coming from rainforests, and the need for the best possible sea defences. Then suddenly, with no clear explanation of the link (none was apparently necessary), the article switches to discussing tourism:

> But while we are trying to protect Eastbourne as a resort, we also need to provide the type of town people want to visit … Holiday bookings are down and there is a genuine problem in attracting people to the town in the light of cheaper foreign holidays and Eastbourne's national image of being a geriatric resort plagued by a polluted sea.[15]
>
> (*Eastbourne Herald*, 25 March 1995, p. 18)

Such a shift only makes sense in the context of a pervasive local understanding of the importance of tourism to Eastbourne's economy, and an economistic understanding of what is important in determining policy; that the Council's policies should be determined primarily by considerations of the local economy.[16]

Global economies

The episode was also economically a global one. At a perhaps banal level, the emergence of a genuinely global economy makes using wood imported from over 4000 miles away possible. Greenheart has been used in sea-defence structures in the UK since the late nineteenth century. Eastbourne's previous groynes, however, were built of oak, of local source. Wood has been imported into Britain from South America since the early period of European colonisation, and a trade existed by the early sixteenth century (see, for example, Dean, 1987). But the scale and purposes of the trade were both much smaller and primarily for luxury, aristocratic uses. It is only from the late nineteenth century and into the twentieth century, with its progressive and radical reductions in the costs of transportation, that a large-scale trade for such purposes as sea defences becomes possible.

One of the specific characteristics of globalisation is that it is an intensification of what Giddens refers to as one of the central features of modernity; 'time–space distanciation' (Giddens, 1985; 1990), or what Harvey (1990) refers to, evoking different imagery, as time–space compression. Modernity 'increasingly tears space away from place by fostering relations between absent others, locationally distant from any given situation of face-to-face interaction' (Giddens, 1990, p. 19, quoted in Saurin, 1994, p. 47). Saurin continues:

> increasingly we produce what we do not consume, and we consume what we do not produce. The separation of material production and consumption also involves the removal and abstraction of knowledge about that self-same relationship. Thus, neither a perfunctory knowledge of the technical procedures of production nor the environmental circumstances and costs of production are visible to the distanced consumer.
>
> (Saurin, 1994, p. 48)

In other words, globalisation makes it possible to make invisible the effects of one's actions on the environment and other societies, because of such distanciation. This is paradoxical, since globalisation simultaneously shrinks the globe culturally, in terms of making us more conscious of events across the globe. Certainly, the inhabitants of Eastbourne and the councillors and engineers involved were aware of the global effects of their actions, and the national media in the UK were able to draw on film-footage of forestry operations in Guyana freely in their stories. Nevertheless, the nature of the experience of deforestation in Guyana by people in Eastbourne is mediated in such a way as to make claims about the sustainability or otherwise of the logging operations opaque to their direct experience (whereas it might not be if locally grown wood were used). They have thus had to rely on evaluating the competing claims of environmentalists and the Council. Council and Posford Duvivier engineers visited DTL's operations in Guyana and claimed to verify DTL's claims at first sight (e.g. in Posford Duvivier, 1994), but their capacity to experience directly other parts of the globe was not available to all participants in the process. At the public meeting in March 1995, Friends of the Earth also produced two people who gave eyewitness accounts of destruction of rainforests in Guyana, claiming to contradict directly the Council's representatives' first-hand experience of DTL's operations, focusing in particular on whether DTL was in fact practising the principles contained in its Green Charter. Protestors also made much of the fact that the trip by Posford Duvivier and Council engineers was paid for by DTL and by Aitken and Howard, the company which would import the timber from Guyana, and that all information produced by Posford Duvivier concerning rates of regrowth of greenheart were themselves supplied to Posford Duvivier by DTL itself, rather than by independent research.[17] But even then, the public was left with a heavily mediated account of processes of environmental change on the other side of the globe but in which they were heavily implicated. They were left largely with a choice of which first-hand account, which expert, to trust. And this choice appears contradictory; we expect the 'compression of the globe' to make such distanced consequences more visible, but in practice they become more distant.

At the other end of the commodity chain, the timber producers in Guyana are inserted into similar global economic networks in different ways, which structured their actions in ways which are perhaps more familiar to students of international environmental politics. As reported by *Channel 4 News*, there was a 'dramatic increase in logging in Guyana during the last five years', as the Guyanan government issued large concessions to timber companies in 'desperate attempts to pay its international debts' (*Channel 4 News*, 11 April 1995).[18] The influence of the debt crisis on deforestation and other environmental problems is well documented (e.g. George, 1992). George shows correlations can be observed between those countries with high levels of indebtedness and with high rates of deforestation. Debt and the structural adjustment programmes imposed by the World Bank and IMF to respond to debt crises create a situation whereby countries become increasingly desperate to find sources of export earnings, and thus create significant incentives for companies and individuals to engage either in logging or in forest-clearance for other land uses, mainly agriculture (ibid.). Other researchers also report correlations between debt and deforestation (e.g. Kahn and McDonald, 1994), although this is disputed by others (e.g. Shafik, 1994).

The policy change is associated with some domestic forces. The year 1989 saw a major change in government in Guyana, as a twenty-year socialist regime came to an end, and was replaced by two successive regimes which pursued neoliberal policies designed to attract foreign investment. Many of the logging companies (including, ultimately, DTL) which were granted concessions to log Guyana's forests were South-East Asian, and the decisions to grant concessions were made by a small body directly under the President's office, whose members are regarded by Colchester (1994) as directly corrupt. But in the background, global forces creating imperatives towards export-led economies are clearly operating in Guyana (ibid.).

How we understand this global connection and dynamic is, however, important. For the Council and its consulting engineers, the connection was seen very clearly in 'problem-solving' terms (Cox, 1986). The underlying dynamics of capitalist globalisation are unproblematic, and can be channelled towards achieving environmental goals. The neoliberal argument that forestry can actively be used to help conserve the forest, to avoid the land it occupies being given over to other uses, is a classic example of this. Posford Duvivier exemplified this argument:

> It should also be remembered that to ban timber cutting in a developing country like Guyana could lead to the destruction of vast areas because the local population would lose sight of the value of it. If they can use the forests for their livelihoods, providing jobs, shelter, fuel and food, they will preserve and care for it. If not, highly destructive operations such as bauxite and gold mining could step in, as has happened in large areas of Brazil.
>
> (Posford Duvivier, 1994)

They go as far as to claim that 'DTL not only needs to ensure the forest is well managed to reassure Western countries, but also to ensure the future livelihood of the local people. Hence it is not in Guyana's or DTL's own interests to strip the forest for a short term profit' (ibid.).

The Council's engineers used a similar argument. In a report to the Council's Environment Committee, they quoted the Overseas Development Administration's Forestry Strategy thus:

> refusing to use tropical timber is not an environmentally or developmentally sound option. It would reduce the long term economic

value of forests and increase the likelihood of their conversion to other land uses. The best ways to help conserve the forests are to work with forest countries to help them manage their forests sustainably, and to maintain a long term market for sustainably produced timber.

(Director of Environmental Services and
Chief Engineer, 1995, p. 4)

Consequently, 'if the recommendations of this report are adopted then…a positive step will have been taken to preserve the Guyanan forests' (ibid., p. 7). Environmental groups involved in principle also agreed with such an approach. Tim Synnott of the FSC, writing in the *Eastbourne Herald* (26 November 1994), suggested that banning the import of tropical hardwoods was not a solution to deforestation, as did Friends of the Earth participants in the debate.

However, there are good reasons to believe this misunderstands the nature of deforestation. At the very least, Posford Duvivier's argument begs many questions. Why precisely does DTL not have an interest in stripping the forest for short-term profit? Why do they have an interest in ensuring 'the future livelihood of the local people'? There is no explanation of any particular characteristics (nor any in other documents produced by Posford Duvivier) of the case which means that DTL does have such common interests either with local inhabitants or in the long-term survival of the forest. It may be that they do have such interests. But this would be arguably rather unusual in comparison with the general role and interests of timber companies operating in tropical rainforests. One of the assertions made in this case by Posford Duvivier to support their claim was that 50 per cent of the income generated by the timber export stayed in Guyana. This claim, however, came from information supplied directly by DTL and Aitken & Howard, which Russell claims that Aitken & Howard could not confirm (Russell, 1995) and which is much higher than the usual figure of approximately 10 per cent given by the International Tropical Timber Organisation, and as cited in communications with Eastbourne Council by Simon Counsell (1995). Such claims would certainly be at odds with the general impact of logging in Guyana. The most extensive account of this is in Colchester (1994), where it is clear that logging, including greenheart, is clearly unsustainable. Even where commitments to some level of forestry which could be claimed to be 'sustainable' are present from the companies involved, there is no government infrastructure to monitor whether such commitments are being acted upon. Colchester also shows persuasively that the priority of the government of Guyana is clearly to maximise logging to earn foreign exchange.[19]

The general conclusion to be drawn from most literature on deforestation is that the scenario painted by Posford Duvivier and Eastbourne Borough Council in relation to the impact of their scheme on forests in Guyana is in theory possible, but in practice unlikely. The patterns of forces underlying deforestation are too complex and convoluted to allow simple statements about the intentions of the company involved, as the effects of these statements of intent will be mediated by government policies (particularly concerning road-building and incentives to convert land to agricultural uses), land ownership patterns, global constraints on the country.[20] All sorts of unintended consequences can be expected which make the impact of logging operations more widely felt than the benign picture painted in Eastbourne.

However, whether or not one accepts this critique of the neoliberal argument concerning the causes of deforestation and appropriate responses to it, it remains the case that such neoliberal arguments literally only make sense within a capitalist economy. The assumption that unless the forest is 'sustainably' forested the land will be converted to other uses assumes a set of capitalist imperatives underlying deforestation, along rational choice lines. People are assumed to be impelled by motives of self-interest (defined in terms of increasing monetary wealth). Alternative lifeworlds are simply assumed to be impossible even to conceive.

Economies of science and sustainability

Sustainability is perhaps the major normative criteria justifying claims about action in relation to environmental politics. Whether or not a particular scheme, project, political or economic ideology is or is not problematic from an environmental point of view, often becomes a question of sustainability. In McManus's terms (citing Blowers, 1993), sustainability is the ultimate 'mobilising concept', the notion around which policies and discourses can be developed regarding the environment (McManus, 1996). Yet the definition of sustainability is itself open to debate. 'Whose sustainability?' is the question underlying this section. Whose version of what sustainability meant in Eastbourne prevailed and why? How did particular versions of what sustainability means serve to legitimise particular courses of action?

A number of competing definitions of sustainability have been in operation in Eastbourne. Coming from central government in the UK, from the Ministry of Agriculture, Fisheries and Food (MAFF), is one which is specifically about the direct effects of projects they fund, and excludes considerations such as the sources of materials used in projects funded by the MAFF. As represented in the Shoreline Management Plan for Eastbourne's stretch of coast,[21] MAFF's definition of sustainability regarding coastal defences is the following:

> A sustainable coastal defence policy is one which provides adequate protection against flooding and erosion in a manner that is technically, environmentally, and economically acceptable, both at the time any associated measures are implemented, and in the future.
>
> (BMT Limited, n.d.)

MAFF's definition is extremely broad. MAFF guidelines for schemes are based on three types of criteria: technical soundness; environmental acceptability; and economic viability and cost-effectiveness. Interestingly, sustainability is included in MAFF guidelines as a technical consideration, not an environmental one. 'Schemes should be sustainable. That is, they should take account of the interrelationships with other defences, developments and processes within a coastal cell or catchment area, and they should avoid as far as possible tying future generations into inflexible and expensive options for defence' (MAFF, 1993, p. 26). In other words sustainability is linked to the emerging holistic discourse of coastal defences (discussed earlier) that projects should take into account their effects further along the coast. But sustainability for the MAFF does not include consideration of the sources of materials used in projects. The considerations of 'environmental acceptability' are also based primarily on local environmental effects (on amenity, or on ecosystems), although there is one mention of 'international obligations' (undefined) in that section (ibid., pp. 26–7). Given that MAFF funded the major part of the project, their definition mattered.[22]

However, the definition of sustainability in public discourse in Eastbourne was rather different. It focused much more closely on the question of the sustainability of extraction of timber from greenheart forests in Guyana. The information pack produced by the Council, for example, engages with a debate over sustainability clearly on these grounds. They state that greenheart is preferable to oak 'on environmental grounds', because it lasts 'up to twice as long as oak', and therefore an additional $6000\,m^3$ of wood would have to be used if oak were the timber of choice (Eastbourne Borough Council, n.d, p. 5). In a section explicitly entitled 'Will Eastbourne's proposals destroy a tropical rainforest?', they discuss the need to get wood only from a 'well-managed' source (see below for more on how slippage between 'well-managed' and 'sustainable' made Eastbourne's proposals easier to legitimise). They cite the neoliberal environmental economics argument already discussed, and invoke the authority of the World Bank and Overseas Development Administration, to suggest that the best way of protecting the forest is to promote 'well-managed' forestry, as this gives the inhabitants of the forest an incentive to manage the forest well, rather than convert it, for example, to farmland (ibid., p. 6). They cite SGS Forestry's certification of DTL's Green Charter and the World Wide Fund for Nature as evidence of the 'well-managed' nature of DTL's operations in Guyana (ibid.).

This version of sustainability was also produced throughout local and national media coverage of the case, and in environmentalist's and engineer's accounts of sustainability. Local press coverage saw the question of sustainability in this way.[23] At the end of the *Channel 4 News* story was a debate about whether the wood to be cut for Eastbourne's sea defences was coming from sustainable sources. This focused on the question of the rates of regrowth of greenheart and whether it was possible to engage in timber production without damaging the ecosystem of the forest as a whole. This became reduced to a question of the rate of regrowth of greenheart. Here the deference to scientists was complete. In the *Eastbourne Herald* (26 November 1994), a debate was staged with protagonists (Posford Duvivier) and antagonists (Friends of the Earth) presenting arguments directly, and the paper included an 'expert' opinion (as if the council's engineers and FoE representatives aren't 'experts'!) from Tim Synnott of the Forestry Stewardship Council (FSC). The FSC is the major accreditor of the sustainability of forestry schemes. The debate concerned whether the 20-year period left by DTL in between cutting (in addition to only taking four trees per two acres in each cutting cycle), and thus on the rates at which greenheart regenerated (defined in terms of m^3 per year) was adequate.

As in relation to the evidence of whether DTL was implementing its Green Charter, participation in this debate was limited to those with the expertise and confidence to use scientific language. The debate was essentially about which scientific evidence was the most persuasive. On the one hand, critics of Posford Duvivier argued that their assessment (that greenheart regenerates at a rate of $1\,m^3$ per year) was based on research from the 1930s and 1950s, and that more recent research should be used. They also said that scientists employed by DTL to evaluate their logging activities were prevented from communicating their work freely. Opponents of the scheme were also able (unusually) to cite a World Bank employee as arguing that a 20-year cutting cycle was too short for greenheart (Russell, 1995). On the other hand, Posford Duvivier and the Council's engineers pointed out the limitations of the research used by FoE and others (that a figure of $0.18\,m^3$ is more appropriate), that this research was carried out in an area of high devastation of a forest, which would show different rates of regrowth to those which would occur in a 'well-managed' forest.

My point here is not to argue about which of these assessments is correct. Clearly, there could be variations in the rates of regrowth depending on a huge number of conditions, and decision-makers have (perhaps arbitrarily) to select a figure in order to make it possible to make a decision. Rather, my purpose is to show two things.

Firstly, the focus on the rates of regrowth question reveals the scientised nature of the discourse. The reduction of the issue of the permissibility of using wood from a particular forest to a question of metres cubed has the effect of excluding all sorts of considerations. Within a scientific discourse, it can be seen as a reductionist act, excluding considerations of the ecological consequences of cutting; the effect on interactions between greenheart and other species in the forest. It also excludes social considerations, of the dependence on the forest of people who use it.[24] In engaging in the debate in this way (rather than in an explicitly ethical manner), the scientists involved arguably legitimised the use of greenheart. It thus serves as an instance of what Beck suggests is the contradiction in science 'between experimental logic and large-scale technological hazards' (1995, p. 161).

Perhaps in a more convoluted fashion, however, 'even those scientists who would be regarded as sympathetic to green interpretations of environmental problems are deeply involved in reproducing the domination of expert systems which are embodied in structures of institutional control' (Smith, 1996, p. 35). Such expert systems are an inevitable part of the growth of an increasingly technologically dominated capitalism, which empowers experts within corporate and state decision-making systems (ibid., p. 30). In Beck's terms, 'protest must accept the basic assumption it intends to contest before it can utter a word' (1995, p. 60). And thus the protestors help to reproduce processes of rationalisation which, as Weber pointed out, are inevitably about the 'disenchantment of the world', or the shift towards understandings of nature as devoid of meaning beyond human instrumental control and use (ibid.). Elvin and Ninghu suggest that one of the consequences of this approach is a 'technological lock-in', whereby the initial intervention in biological processes creates the need continually to intervene later, but the consequence of this is 'not as it sometimes is – overall – disaster, but something much subtler, – conservatism' (1995, p. 48).

The second point, however, is that much of this is a question of political economy. It is a question of the power of businesses to use scientific definitions of sustainability and slippage between the precise language of scientists and the looser language of public discourse to justify actions it appears clear that they had always intended to take, with or without expert opinion on the sustainability of the project. Science served to legitimise deforestation. As debates are reduced to questions of which expert to trust, those who become the primary translators of scientific information for political decision-makers (in this case, Posford Duvivier) are privileged in their interpretation or appropriation of expert knowledge. The *status quo* is the evidence given by Posford Duvivier, and objectors (FoE and others) have to disprove such evidence. When the debate is reduced to the trading of scientific papers, disproving such evidence becomes difficult, when both many of the persuaders themselves are laypeople, or scientists coming from other disciplines, and in particular those making final decisions (the councillors) are laypeople.[25]

However, which scientific evidence was chosen seems not unconnected to how that evidence supported other aspects of the interests of the actors involved. The various actors can be seen to have used versions of sustainability which were consistent with economic interests, or constructed alliances with economic groups in order to strengthen their political hand in relation to how they wanted sustainability to be constructed. There was a noticeable slippage in usage between the term 'well-managed', which was the precise term used fairly consistently by SGS Forestry, Posford Duvivier, and the council's engineers when they were narrowly describing DTL's logging activities, and the term 'sustainability', when the council (particularly councillors, although also sometimes the engineers) and Posford Duvivier were making more general statements of intent regarding timber extraction, or when the local media were presenting the arguments in print. This slippage helped Posford Duvivier and the Council's engineers legitimise what had been, since March 1993, the 'preferred option' (Eastbourne Borough Council, 1993, p. 743).

This point is perhaps made more starkly by the relationship between Posford Duvivier and the timber companies. The fact that Posford

Duvivier's evidence about the sustainability of greenheart (in terms of its rates of regrowth) came directly from DTL and Aitken & Howard was much used by oppositional groups, suggesting that such evidence cannot be regarded as 'reliable'. Clearly, if it was the case, as FoE and others alleged, that Posford Duvivier had decided on using greenheart on engineering grounds (or as hinted at occasionally more strongly, perhaps themselves had a vested interest in using greenheart), but needed to legitimise its use, then using evidence of rates of growth consistent with DTL's logging practice would be more congenial to them than using evidence which suggested that such logging was exceeding the capacity of the forests to grow back. FoE and local timber companies alleged that the specifications for tendering had been written in order to exclude oak from consideration, and that Posford Duvivier had reported erroneously to the Council that there were insufficient supplies of oak locally, without consulting potential suppliers.

However, the way that oppositional groups used sustainability also had a political economy, albeit perhaps less direct. Michael Le Page, the local Friends of the Earth activist most involved in its campaign, made this explicit. 'Chopping down virgin rainforest cannot be justified, especially when it puts British jobs at risk', he said, during the campaign (quoted in *Eastbourne Herald*, 4 March 1995). FoE actively made alliances with local timber firms, notably Timber Management Sawmills Ltd, who claimed to be able to produce, sustainably, sufficient oak from local sources, for the scheme, and helped to try to defend the interests of those companies. They were supported in this by East Sussex County Council, who later argued in evidence to the House of Commons Agriculture Committee that there were over 750 000 tonnes of oak which could be produced locally, at sustainable rates, from woods in East Sussex (Agriculture Committee, 1998b, p. 73). Thus again, given this background, FoE developed an interest in showing that DTL's rates of logging were unsustainable, and supplied this information to the local timber firms who used it in their own evidence to the Council.

So while the politics of deforestation in Guyana was scientised, this occurred in such a way as to benefit particular actors over others. Those already legitimised by the Council to act on their behalf and advise them became the legitimate providers of knowledge about, for example, rates of regrowth of greenheart trees. While sometimes a scientisation of environmental problems may produce ideological resources to oppose environmental destruction, in this case it clearly had the opposite effect. Environmentalists were arguing that sustainability required that the cutting cycles should be greatly increased, whereas the timber countries and companies, the Council and its consultants, and the Ministry of Agriculture, Fisheries and Food, were all happy that cutting could continue until scientific evidence suggested otherwise. Science, having reduced the world to a series of experiments, meant that such experiments were legitimate until proved otherwise. Arguing on this basis, environmentalists arguably had little chance of persuading decision-makers who depend heavily on their own officers for their primary source of advice, and necessarily build up relations of trust with those officers. It is structurally difficult for outsiders to overturn decisions essentially already taken. As Smith points out, when environmentalists have to 'play the game', while they may contribute to improving the environment in some ways, they inevitably 'play' by the rules which 'requires the representation of the issues within a form that is acceptable, and understandable, within the structure of the decision-making process itself. This is presented primarily in terms of expert knowledge ... ' (Smith, 1996, p. 37). Thus environmentalists help to embed processes of rationalisation of the environment which, in Smith's terms, are 'at the heart of the problem' (ibid., p. 38).

Conclusions

This chapter has tried to show how decisions concerning the environment are always also about the reproduction of various forms of social power. The decisions in Eastbourne reflect the way that various actors are embedded in local and global economic systems which structure their actions heavily. Broader discursive shifts concerning sea defences do not undermine the way in which such projects have always been about nation-building; rather, they reflect a shift in the content of the nations being built. The scientised nature of environmental decision-making acts to confer authority on certain actors and exclude or marginalise others.

The cumulative effect of the structuring of environmental decision-making is that political institutions should be regarded as fundamentally ecologically problematic. Their overriding priorities are to engage in practices which are unsustainable in ecological terms. This therefore undermines the assumptions of liberal institutionalists that such institutions are neutral with respect to environmental change. I move now to examine the dynamics of broader social practices which generate such environmental change.

5
Car Trouble

Introduction

An advert for the Nissan Primera in 1996–97 has a man get up in the morning, dress, leave the house, drive to work, get to work, which turns out to be back at home. As Eagar puts it 'A smug yuppy in bed says "I think I'll drive to work today, Mrs Jones." His wife replies: "Fine, Mr Jones." ... You see, they work from home, but he likes the car so much, this Nissan Primera, that he still commutes' (Eagar, 1997). The punch line of the ad is 'It's a driver's car, so drive it'. When asked the reason for such a slogan, the writer of the ad, David Woods, said that '"We were looking for an unnecessary reason to drive a car"' (quoted in ibid.).

At the same time, British politics produced a most unlikely hero, known as Swampy. For a few weeks in January and February of 1997, Swampy was one of the most talked-about figures in British political debate, and his popularity endured throughout early 1997. Swampy was one of five protestors against a road-building scheme on the A30 in Devon who managed to get themselves down an extensive network of tunnels they and others had constructed when their protest camp, Fairmile, was evicted. Swampy was the last of the tunnelling protestors to be pulled out of the ground by 'rescuers' employed by the bailiffs to clear the way for the road, seven days after the five had gone underground. Although a certain proportion of the media construction was negative, portraying the protestors as 'evil scum', unemployed outsider activists rejecting the values of 'normal' people,[1] what was interesting about this case was that in most of the British press and TV discourse, the coverage was largely positive. The predominant image was constructed though discourses of youthful active heroic idealism, and the protestors were normalised through various means. This mostly involved discourses of family, class, and nation, whereby the road protestors (who in earlier episodes had been constructed as deviant outsiders in the media) were now constructed as representative of the values of 'middle England'. 'Beneath the encrusted grime and matted hair beats a respectable suburban middle-class heart', wrote the (conservative) *Daily Express* about Swampy (3 February 1997, p. 10).[2]

The normalisation of those willing to break the law, not to say risk their own lives, to prevent road-building and challenge car culture, says much about how deep those challenges to car culture have gone. The depth of such challenges in public discourse can perhaps be seen when the conservative *Spectator* magazine has a column entitled 'Not Motoring' (*Independent*, 27 January 1997, p. 20).[3]

In 1984, and republished in translation into English in 1992, Wolfgang Sachs wrote perceptively that 'The problem with the automobile today consists precisely in the fact that the automobile is *not* a problem' (1992a, p. vii). But the processes just briefly described concern the way in which spaces where the car has become 'a problem' have been created. The car (and road-building, always to further automobility) has been undergoing particular challenges in the UK in the 1990s, involving a challenge to the presumption in favour of the motorist in suggesting that driving should be only for necessity and avoided if possible. It is therefore perhaps only reasonable for car manufacturers, and the manufacturers of car culture, to hit back with an advocacy of the virtue of driving simply for its own sake. It is at points where hegemonic ideologies are under threat that they need to reconstitute their power where possible. Of course, there have always been challenges to the car since its inception (see below), but what is interesting about the contrast presented here is that it is at the juncture at which mainstream acceptance of the points made by protestors and (perhaps less so) the values they espouse is forthcoming that the value of ads espousing 'unnecessary' driving becomes salient.

As car culture and Margaret Thatcher's (in)famous 'great car economy' are under threat, advertisers and other manufacturers of car culture shift the discursive frame within which car culture has to be reproduced. The Primera advert is one such strategic shift. This strategy is one where car culture is brazenly espoused – the joy of driving for its own sake is celebrated, as the problematisation of the car since the mid-late 1980s is discursively erased. In a later version of the same series of Primera ads, an environmental connection is at least implicit. In this one, a Florida TV weatherman announces that a particular hurricane has switched course and is headed for downtown Miami.

He gives instructions so that people should under no conditions leave where they presently are. The ad then cuts to him driving his Nissan Primera down a Miami highway singing along to The Troggs' 'Wild Thing', with the 'It's a driver's car, so drive it' slogan rounding off the ad. Here, the prevalent image of motoring being about the open road without other car drivers present is simultaneously revalorised and satirised – like the Nissan Almera ad (see below), a nostalgia for a past where the car could be presented unproblematically is produced (perhaps reinforced by the choice of music), knowing that it can no longer be portrayed in such a fashion. Also, as in the case of Eastbourne's sea defences, the irony of promoting something which causes environmental changes which can then be used to promote further consumption of those products, is hinted at. Increases in severity and frequency of hurricanes are among some of the secondary effects thought by many scientists to result from global warming, itself caused by, among other things, cars.[4]

Other strategies are, however, adopted by the promoters of car culture. For example, Baird (1998, pp. 152–3) suggests that images of the open road are being dropped, as the advertisers recognise that people mistrust such images as they are increasingly discordant with peoples' everyday experiences of traffic jams. She cites ads such as one for the Vauxhall Vectra, which is located in a traffic jam. Where other drivers get furious with frustration, the driver in the Vectra is 'cool and relaxed' (ibid., p. 152) because of the Vectra's air-conditioning.

One predominant alternative strategy is to restructure the car's legitimating motifs to take into account challenges to the car's dominance from environmentalism. Two adverts running alongside the Primera advert adopt such a strategy; that of the Ford Ka, and the Honda Dream II, as do more recent ones, that of the Renault Laguna, and of the Vauxhall Vectra.

In the first of these (actually a series of TV ads), the car's body is in places almost invisible. In the first of the series of ads, the car only appears fleetingly at the end of the advert, being submerged in images of 'nature'. In others, it is presented through a particular configuration of the themes of nature, technology and progress, where technology and progress are presented in terms of harmony with nature (to the point of literal invisibility within nature's emblems, often trees in these ads). The Honda ad is perhaps less subtle, certainly less surreal, in portraying Honda's prototype solar car as the car of the future, where technology is again placed at the service of progress in addressing questions of environmental change.

The ad for the Renault Laguna, running in 1998, has well known (socio)biologist Steve Jones delivering a lecture on evolution and natural selection. In the background earlier models of Renault cars 'evolve' into the current Laguna model. Jones intones that evolution is 'not just a theory, it is going on all around us'. The appropriation of nature as evolution serves to represent the car in question as being refined and improved progressively, adapting to its environment as would an organism (or gene). At the same time, this is a classic example of the processes discussed by Haraway (1991) whereby social norms are read on to nature, and then back on to society as naturalising justifications for social practices. Discourses of evolution and natural selection read modernist notions of competition, selection, adaptation and most importantly progress on to nature, and then back, in this case, on to car design. Such a move renders car-centred development 'natural', and by implication irrational, if not literally impossible, to resist.

In a different vein, Vauxhall ran a TV ad as part of their campaign for the Vectra in 1998, which had as its general slogan 'for the corners of the earth', focusing on how well the Vectra takes corners while implying how the owner will want to drive everywhere in it. In one of this series, the advert starts in a road protest which aims to prevent a forest being cut down to make way for a road. The roadworks would have straightened out a road which currently wound around the forest. The protest succeeds in preventing the road being built and then one of the protestors (conspicuously in short hair, chinos and polo shirt, as opposed to dreadlocks, combat trousers, etc.) leaves the group and the ad cuts to him driving around the forest on the old road, happy with the winding road because of how well his Vectra 'corners'. To top this audacious appropriation of the road protests with a two fingered salute, the driver passes one of his fellow ex-protestors who is now hitchhiking, and shouts out of his window 'Get a job'.

Wernick (1991) suggests that the direct dealing with ways in which the car's relationship with the environment has been problematic is one of the two themes which have been central in the ways that car advertising has changed since the early 1970s (the other concerns the family, gender and patriarchy). Car ads have increasingly had to take into account the environment and critiques of techno-optimism more generally, and have tended to reappropriate nature for the purposes of the car (Wernick, 1991, pp. 77–9).

One final point which Wernick makes is that another strategy in dealing with these challenges to car culture is to use nostalgia, which he suggests has become a prominent theme in car advertising. He discusses at length a (late 1980s) Vauxhall Cavalier advert in which a stylised advert from the 1950s is shown, with Dan Dare/Buck Rogers style futuristic cityscape, a nuclear family in its car with the father at the wheel, unproblematically taking his family 'for a spin'. The advert shows all the features which cars in the future might have. On the opposite page of the ad is a photograph of the Vauxhall Cavalier against a backdrop of grass and trees, and some (unspecified) industrial (but again futuristic in design) building in the background. There is no family or people present, but nature is represented. The caption reads 'Who said tomorrow never comes?' The point of nostalgia is to concede the ground that the car can no longer be conceived of unproblematically, but to encourage people to hark back to an age when it could be. In the Cavalier ad, 'tomorrow's-car-of-the-future-today is presented as fulfilling a technological dream which in crucial respects, the promotion for it also disavows' (ibid., p. 88).

In 1997, the ad for the Nissan Almera, another prominent advert, also uses such a strategy. This is a pastiche of the 1970s action detective series *The Professionals*,[5] similar in format and style to the American *Starsky and Hutch*. The advert involves much screeching of car-tyres, driving through puddles and market barrows full of fruit, shouting, and so on. The billboard version is shot in blue-black and white. Although 'nature' or the environment is nowhere present in this advert, it clearly fits with Wernick's argument that nostalgia as a discursive strategy responds to challenges to car culture (see also Baird, 1998, pp. 141–2 on this ad).

Car culture, and a car-based economy, therefore do not simply exist 'naturally'. They need to be reproduced and legitimised. They often come under challenge and must therefore reinvent themselves to adapt and take advantage of new discursive shifts. Nevertheless, the car is now globally *the* dominant mode of transport. The world's towns and cities have been totally restructured to serve the need of car-centred transport systems. Many people's lives are now literally dependent on cars for meeting basic needs. Freund and Martin (1993) term this 'autohegemony'. How did we get here?

Autohegemony[6]

There is a naturalistic tendency in much writing on the rise of the car, particularly among economists or business historians. This tendency explains the rise of the car in terms of the natural advantages it has over other forms of transport and the way it taps into powerful forces in human psychology. Business studies writers Moxton and Wormald come up with one of the more bizarrely psychologistic of such explanations:

> The truth is that our attachment to cars is profoundly rooted – not only in the practical necessities of life but also in our emotions. Research shows that there is a deep psychic connection between freedom and movement. Babies achieve locomotion. Adults re-experience it through the motor car. Waiting for a bus or a train unleashes hidden, unconscious fears of abandonment in many.
>
> (Moxton and Wormald, 1995, p. 33)

Overy also tends to naturalise the growth of the car, arguing that 'the reception and rapid evolution of the motor vehicle … needs little explanation' (1990, p. 57). The way writers often discuss the (usually American) 'love affair' with the car reinforces such naturalistic notions (e.g. Davies, 1975, p. 7; Flink, 1975, ch. 1).

These accounts of the rise of the car are misleading and highly ahistorical. Much of the rise of the car can be explained in terms of political-economic forces (narrowly understood). For example, cars have widely been seen to have played a fundamental part in the promotion of economic growth in the twentieth century. Proponents and social critics alike argue that the car has been central to promoting growth. This has been firstly because of the way in which investment in the car industry directly stimulated a whole host of other industries (petrochemicals, steel, engineering, road-building) and thus the economy as a whole. Secondly, cars enabled a faster and more flexible means of distributing goods through the economy than previous transport modes, thus accelerating growth. Finally, the car industry innovated extensively in production techniques, producing the broad political-economic shift known often as 'Fordism', which when adopted throughout the economy, produced substantial increases in industrial productivity. This has therefore been central in legitimising the car's expansion, enabling the car to become perhaps *the* symbol of progress for most of the twentieth century.[7]

Given the state's structural role in promoting accumulation, it is no surprise that once the car's potential in accelerating accumulation was realised, states began to promote the car vigorously. The car industry offered significant improvements in the capability to commodify means of mobility, and at the same time accelerate the movement of goods and people in the economy. Promoting the car through hidden and not-so-hidden means has helped it to become the dominant force it has. Such state promotion of cars is perhaps best understood in terms of the state's structural role in capitalist societies, its general imperative to support the conditions for capital accumulation (e.g. Jessop, 1990).

The promotion of the car economy by the state has had perhaps four main facets. The first of these has been road-building (both within and between urban areas), which has increasingly meant that, since road-building is almost always paid for out of general taxation (while investment in other transport means is not), this is a subsidy to car-users not given to other transport-users. The second is the progressive neglect and downgrading of public transport and non-motorised forms of transport (e.g. Wolf, 1996, pp. 75–81, 117–23). Thirdly, various fiscal measures effectively subsidise car-use relative to other forms of transport, such as tax relief on the provision of company cars. Athanasiou estimates that the total annual value of subsidies to the car in the US alone are approximately $400bn (1996, p. 264). Finally, states have occasionally colluded with car manufacturers to remove competition to the car, most famously in the case of National City Lines, a company established by General Motors, Standard Oil of California and Firestone Tire Company, which, with the active consent of the city governments concerned, systematically bought up and dismantled tram lines throughout the US.[8]

Modernity, speed and identity

I focus on the international political economy of the car as briefly summarised above in Paterson (2000a). Here I want to focus on the way in which the emergence of a car *culture* has been crucial to establishing the car's prominence. It would be inadequate and misleading to characterise the rise of the car simply as a story of economic manipulation and political promotion. What Gorz (1980) called 'the social ideology of the motorcar' is deeply entrenched in individual and collective identities. Such an ideology has been able to become so deep-rooted because of the way its manufacturers have been able to link it to pervasive ideologies and widely valorised themes. 'The alliance of those whose livelihood depends on a robust automobile-centred culture ... also feed on the culture symbolism of the automobile: freedom, individualism, mobility, speed, power, and privacy' (Tiles and Oberdiek, 1995, p. 137). Such symbolic connections are ubiquitous in the variety of cultural arenas, from pop music, to film, to twentieth-century novels, where cars are widely valorised (Baird, 1998, pp. 28–35).

Of these six themes of 'freedom, individualism, mobility, speed, power, and privacy', I will focus on the notions of speed and mobility. In many accounts, speed, and acceleration in particular, are taken as perhaps the primary feature of twentieth-century modernity. In Freund and Martin's terms, 'Speed *is* the premier cultural icon of modern societies' (1993, p. 89). Berman makes this explicit in his account of modernity. While for him the central general feature of modernity is that 'all that is solid melts into air', in the twentieth century this continual change is effected through continuous acceleration brought about by new transportation technologies, primarily the car. He quotes Giedion, Le Corbusier's most famous 'disciple' (Berman's term) in architecture and urban design, relating this to the car explicitly. 'The space–time feeling of our period can seldom be felt so keenly as when driving', Giedion wrote in 1938–39 (Giedion, 1949, quoted in Berman, 1982, p. 302). Berman also quotes Le Corbusier thus:

> On that 1st of October, 1924, I was assisting in the titanic rebirth of a new phenomenon ... traffic. Cars, cars, fast, fast! One is seized , filled with enthusiasm, with joy ... the joy of power.
>
> (Berman, 1982, p. 166)

The association is then one from modernity and modernisation, with notions of progress built in (which are so valorised that they cannot be resisted) with acceleration and increasing speed, and thus the car becomes a primary symbol of modernity itself. The driving experience becomes itself an end, not simply a means. As a US car ad in 1993 suggests, 'Illogical as it may seem, the simple act of motoring down the boulevard, exhaust burbling, that's what Viper ownership is all about. Only behind the wheel does it all make perfect sense' (quoted in Freund and Martin, 1993, p. 3).

Speed has been one of the main motifs underlying popular constructions of the car. At times, it has been a central part of advertising strategies, with focus, for example, on the time taken to accelerate from 0 mph to 60 mph. In the UK at least, such a focus in adverts is now against the Advertising Standards Authority's (an industry self-regulating body) code of conduct because of safety concerns, but outside formal advertising, speed is still highly valorised. A programme like *Top Gear*, a highly popular TV programme about cars, focuses heavily on speed in the way it portrays cars. Its most prominent recent presenter, Jeremy Clarkson, is well known for glamourising the speed of cars demonstrated in the programme. In one episode of *Top Gear*, he drove a Jaguar XJR above 100 mph, proclaiming that it was 'bonkers fast ... rockets from nought to 60 [mph] in five seconds' (as quoted in Baird, 1998, p. 187).

Thrift also makes speed a central feature in his account of changes in contemporary societies (1996). Although he dislikes the term 'modernity', he groups together three themes – speed, light and power – under the collective term 'mobility', to characterise such changes in ways similar to Berman's account of modernity. Thrift suggests that the three 'have been crystallised by considerations of a commonplace, even banal, image; an urban landscape at night through which runs a river of headlight' (ibid., p. 257).

Like Berman, Thrift focuses on nineteenth- and twentieth-century change in terms of the mobility produced by transport and communication technologies. In the nineteenth century, the consequences of the adoption of such technologies (I pick up only on his accounts of such consequences in terms of speed and transport here) were fourfold. Firstly, they produced a 'change in the consciousness of time and space', involving increased attention paid by people to smaller distinctions in time (leading to the development of a market for watches), the emergence of travelling to work as a social practice, and the increasing experience of landscape from a moving rather than stationary vantage-point (Thrift, 1996, p. 265). Secondly, there was the way that literary texts paid attention to speed either in terms of celebration of machine-driven acceleration or protests against the increasingly hurried nature of life (ibid.). Thirdly, there was a change in the nature of subjectivity, involving an 'increasing sense of the body as an anonymised parcel of flesh which is shunted from place to place' (ibid., p. 266). Finally, prevalent social metaphors emerged reflecting the preoccupation with speed, notably 'circulation' and 'progress'. Speed thus culturally became understood as causally connected to progress. 'Whatever was part of circulation was regarded as healthy, progressive, constructive; all that was detached from circulation, on the other hand, appeared, diseased, medieval, subversive, threatening' (Schivelsbusch, 1986, p. 195, quoted in Thrift, 1996, p. 266). Travel thus became an end in itself (Thrift, 1996, p. 267). By the late twentieth century, such a conception became deeply embedded – 'Travel is now thought to occupy 40 per cent of available "free time"' (Urry, 1990, p. 6, as quoted in Thrift, 1996, p. 280).

In the twentieth century, Thrift suggests that speed continued as a predominant social theme, but that through at least to the 1960s, there was little qualitatively different about the consciousness of space and time from the nineteenth century (Thrift, 1996, p. 277). The car emerged, however, as the 'most important device', the 'avatar of mobility' (ibid., p. 272), which has helped to entrench the notion of a society where the 'only truly profound pleasure [is] that of keeping on the move' (ibid., p. 273, quoting Baudrillard, 1988, pp. 52–3).

But the importance of the way cars could be linked to freedom through notions of speed and mobility has also had a context in the way that industrial societies have become increasingly regimented and bureaucratised to serve the needs of industrial production. 'More than any other consumer good the motor car provided fantasies of status, freedom and escape from the constraints of a highly disciplined urban, industrial order' (McShane, 1994, p. 148; also Ling, 1990, pp. 4–5). This fantasy was particularly important since even in the US, the most motorised country in the world, no one was actually able to commute or even, apart from a very tiny number, travel far (on holiday, for instance), until after the Second World War (ibid., pp. 125–7).

The association of cars with speed, and with speed as their main legitimising motif, is widely recognised in contemporary accounts (e.g. Ross, 1995, p. 21; McShane, 1994, p. 114; Wolf, 1996, ch. 13). This can perhaps be best seen in the way that many critiques of the car focus on the notion of speed and mobility. They focus on the fact that in many cities, average speeds are now often no faster than they were before the car's emergence, with London averaging 7 mph, Tokyo 12 mph, and Paris 17 mph in the rush hour (Wajcman, 1991, p. 127). The critiques in popular accounts, such as Ben Elton's *Gridlock* (1991) focus, among other things, on the irony and frustration produced by the car. The myth of the car is centred on speed and mobility, but since it simultaneously produces congestion on a scale never previously seen, it is simple (at least in Elton's fictional account) for political elites to manipulate this and produce a total Gridlock of London, in order to justify further road-building.[9]

A classic critique of this is in Ivan Illich's *Energy and Equity* (1974). Illich calculated that:

> the typical American male devotes more than 1,600 hours a year to his car. He sits in it while it goes and while it stands idling. He parks it and searches for it. He earns the money to put down on it and to meet the monthly installments. He works to pay for petrol, tolls, insurance, taxes, and tickets. He spends four of his sixteen waking hours on the road or gathering his resources for it.
>
> (ibid., p. 30)

On this basis, Wolf suggests that speed is a myth, not in the sense of an organising social motif, but as something false and to be debunked. He calculates that taking all these factors which Illich discusses, the 'real' speed of car transport averages at approximately 20 kmh, about the same speed travelled by a 'very fit cyclist' (1996, p. 187).

The modernisation of which the car was the ultimate expression is therefore deeply embedded in individual identities. However, the way this is embedded in those identities seems to me not best expressed in Gorz's terms. For Gorz (1980) ideology is used in the sense of something which *masks* reality. All that is required is to unmask this ideological cloak and social change becomes possible. Similarly, Wolf (1996) and Gartman (1994) both treat the way that cars are embedded in identities as primarily a psychological reaction to alienation in the capitalist labour process; a means by which capitalism displaces the alienation it inevitably produces. The car for Wolf is then a 'substitute satisfaction' (ibid., p. 192), or for Gartman, an 'ersatz satisfaction' for the degradation of work under Fordist mass production (1994, p. 12). But the notion of false consciousness which underlies these interpretations is deeply problematic. While not wishing to dispute the 'facts' they present (Gorz's argument about the impossibility of everyone owning a car, Wolf's concerning the myth of speed, both drawing on Illich), it seems to me more useful to take seriously the reality and depth of the identities produced around the car. They should not be dismissed as false consciousness, but should be understood as deeply embedded. As Gartman argues, 'rather than see the needs appealed to by consumer goods as false needs engineered by the culture industry, my formulation conceptualizes them as true needs for self-determining activity channelled by class conflict into the only path compatible with capitalism – commodity consumption' (1994, p. 11). However, Gartman still relies on viewing mass consumption, notably of the car, as a displacement from the alienation produced by capitalist mass production.

Berman again seems to me to understand the relationship and contradiction here better;

> This strategy [of the promoters of the 'expressway world'] was effective because, in fact, the vast majority of modern men and women do not want to resist modernity: they feel its excitement and believe in its promise, even when they find themselves in its way.
>
> (1982, p. 313)

As Berman quotes Allen Ginsberg, the forms of identity produced in this process are not false, imposed purely to meet someone else's interests; they are more like 'Moloch, who entered my soul early'

(1982, p. 291). The car is partly constitutive of who it is to be us, not something externally imposed on us through deceit. Understanding the relationship in terms of notions of the cyborg developed in general by Haraway (1991), and invoked in relation to the car by Thrift (1996) or Luke (1996), for example, gets closer to the complexities of the relationship between human identities and the machines through which such identities are shaped. The transformation of those identities cannot be achieved by simply showing their 'false' nature.

Cars and social inequalities

It should also be remembered that modern societies are highly unequal, and arguably depend on such inequalities for their continued reproduction. While the car's success owes much to its symbolic resonance with predominant themes in twentieth-century modernity (and the capacity of its manufacturers and promoters both to present the car in such a way, and to promote a particular form of modernity with which the car is resonant), the impacts of the car are also highly unequal. Access to cars is greatly affected by gender, race and class. The socio-spatial consequences of car-centred development (most prominently suburbanisation) have also helped to reproduce such social divisions. Entrenched understanding of the identities surrounding cars (and of the cars themselves), particularly in terms of gender, serve to reproduce such inequalities at the deep level of individual and collective identities.

While initially clearly a high-status object for the very affluent, very early on the car was appropriated as something which could democratise society and erase class barriers. Henry Ford was very clear that this was the purpose of the Model T Ford, and that mass ownership of cars was a direct substitute for class politics:

> the time will not be far when our own workers will buy automobiles from us. ... I'm not saying our workers will ... govern the state. No, we can leave such ravings to the European socialists. But our workers *will* buy automobiles.
>
> (Quoted in Wolf, 1996, p. 72)

So the car has been sold as a democratising force. Of course, this was also used by non-democratic political leaders. Hitler in particular promoted the *ideology* of mass motorisation (no actual German civilians drove Volksmobiles until after the Second World War [Wolf, 1996, pp. 98–9]) and the first large-scale motorway construction projects. In

Hitler's case, it had distinctly military purposes lying behind it – the design for the Volksmobile Beetle was such that it could be put to both civilian and military uses, and the motorways were constructed to make troop movement quicker and more flexible than could be achieved by rail transport. But as an ideology, it was also designed, as it was in Ford's rhetoric, to help erase class differences and 'to meld the German people into unity' (the very name 'people's car' was part of this) (Sachs, 1992a, p. 53; Wolf, 1996, pp. 97–101). The Nazis used the metaphor of 'circulation' to promote motorisation, connecting car use to the ideology of blood and soil (Sachs, 1992a, pp. 47–50).

This rhetoric of the car as a democratising force endures. Overy, discussing the democratic pretensions of Ford and other car manufacturers, tends to reproduce their claims uncritically. 'There is a very real sense in which the democratization of motor-car ownership and motor transport matched the corresponding political shifts towards mass politics and greater equality' (1990, p. 62). But gender, racial and class inequalities have been built into the promotion of cars, and perhaps into their nature as a technology. A car-centred economy has helped to reproduce such inequalities, themselves crucial in the reproduction of capitalist societies.

Cars are gendered in a number of ways. The most commonly observed is in the imagery of cars produced in adverts but also in the design of cars itself (Wernick, 1991, pp. 72–5; Wolf, 1996, pp. 207–8; Freund and Martin, 1993, pp. 90–3; Wajcman, 1991; McShane, 1994, pp. 132–40). Cars are either produced as masculine, 'figured as rocket, bullet, or gun, that is as a sexual extension of the male', or as 'Woman … as flashy possession, mistress or wife' (Wernick, 1991, p. 74). Either way, prevailing patriarchal constructions of masculinity as dominance (where the car simply becomes an extension of the man) and femininity as submission ('she handles really well') are reinforced. Wajcman (1991, p. 134) suggests the latter construction is dominant. 'Manufacturers encourage the male user to perceive his machine as a temperamental woman who needs to be regularly maintained and pampered for high performance' (Chambers, 1987, p. 308, cited in ibid.). More prosaically, car ads are predominantly aimed at men (although this is gradually shifting as the proportion of car-buyers who are women increases) and use sex to sell cars (e.g. Baird, 1998, pp. 147–8; Marsh and Collett, 1986). Even where cars are marketed at women, sexist portrayals of women in them are still commonplace (despite a common liberal narrative of progress in eliminating sexism from public culture), as in the recent (1998) use in the UK of 'supermodel' Claudia Schiffer disrobing before getting in her Citroen Xsara.

Occasionally, the car is presented as a liberator of women. In a survey of the car, *The Economist* emphasises such a claim (25 January 1986; see Tiles and Oberdiek, 1995, p. 135). Virginia Scharff (1991) gives the fullest account of the rise of the car in such terms. Analysing women's relationship to the car in early twentieth-century America, she argues that the car made it possible for women to lead more independent lives, engage in a broader variety of work, break down established norms of femininity, as well as facilitating the organisation of the suffrage movement in rural areas. Apart from the consequences of the car for women (see below) against which such benefits have to be weighed, a fatal weakness of Scharff's account is that the women who appear in her narrative are almost exclusively inordinately privileged. They come from the ranks not even just of the (upper) middle classes, but the extremely wealthy. Generalising on the basis of their experience seems particularly unwise.[10]

For most writers on the subject, cars have been important technologies in buttressing male power. Connell argues that cars have been central technologies in tying working-class men to a notion of hegemonic masculinity which maintains male power in technological capitalist states, giving notions of masculinity tied to aggression, violence and technology a mass base (1987, p. 109–10). 'The gradual displacement of other transport systems by this uniquely violent and environmentally destructive technology is both a means and a measure of the tacit alliance between the state and corporate elite and working-class hegemonic masculinity' (1987, p. 110). In a similar vein, McShane shows how the emergence of the car occurred at a time when industrialisation was eroding traditional forms of masculinity, and that 'the motor car served as a battlefield in the wars over gender roles that were so important in early-twentieth-century America' (1994, p. 149; cf. Scharff, 1991). In the US, the late nineteenth century experienced a moral panic surrounding masculinity, as industrialisation meant many men were no longer using their 'brute strength' in their daily lives, and women were making important inroads into many occupations. Masculinity was thus refigured to include notions of 'mechanical ability', in an attempt to maintain particular workspaces as male domains. 'Mechanical ability was becoming an attribute of gender, not social class' (McShane, 1994, p. 153).

While cars have therefore been important symbolically and materially in reproducing male power, they have also been directly instrumental in the way that public space has been (re)masculinised in the twentieth century. While cities and towns have been progressively more organised spatially to serve the needs of car-centred transport, the diversity of uses of public space has declined. Simultaneously, men have been able to make cars predominantly theirs. Wolf (1991, pp. 204–5) gives figures for Germany regarding differential access to cars by men and women. There, 79 per cent of eligible men have a driving licence, while only 50 per cent of women have one. Forty-seven per cent of German men have 'continuous' access to a car, while only 29 per cent of women do. Only 22 per cent of the cars registered in Germany are owned by women.

Public spaces in central cities have been, therefore, increasingly masculinised, and one of the predominant attitudes to city centres expressed by women is now fear (Wolf, 1996, pp. 206–7; Wajcman, 1991, p. 131; Tiles and Oberdiek, 1995, p. 136). The process of suburbanisation has in many cases involved direct decreases in the mobility of women (at least those in 'traditional' nuclear families with male breadwinners), moved out to homes in suburbs, away from friends, shops, and so on. And even where distances remain walkable, women's mobility has often been hit by urban road-building as dual carriageways and urban freeways make walking across urban areas more difficult, dangerous and/or time-consuming (Kramarae, 1988, p. 121, cited in Tiles and Oberdiek, 1995, p. 136).[11]

Such spatial reorganisation has occurred predominantly using social divisions around race and class. Suburbanisation was founded initially on American rural anti-city ideology, precisely at a time when US cities were becoming more concentrated and when the ethnic mix of the cities was becoming more diverse. Suburbanisation allowed middle-class white Americans the chance to revitalise a rural ideal and escape from the working classes and those from other ethnic backgrounds to maintain their privileged status (McShane, 1994, p. 123; Freund and Martin, 1993, pp. 103–4). Winner's (1980) analysis of the New York parkways built by Robert Moses is again instructive here. Moses was explicit in creating white middle-class spaces in the parkways, by engineering-out public transport predominantly used by black and working-class Americans.

Class has operated in a more complex fashion, however, in relation to the car. As shown already, an ideology of cars promoting a blurring of class boundaries persists. But access to cars is highly differential. We have already seen this regarding gender-based inequalities, but it also applies to other forms of inequality. Just as one example, in the UK, in the richest 10 per cent of the population, 90 per cent of households have cars, while in the poorest 10 per cent, only 10 per cent have cars (Hamer, 1987, p. 2). Moreover, the car-centred economy has *reduced* mobility for many; those dependent on public transport or cyclists and pedestrians, and those relocated by the suburbanisation produced by the car, but without access to one.

Cars and environmental change

This argument implies that responding to environmental change and moving towards a more sustainable society necessarily involves a shift away from a car-based economy. It is perhaps worthwhile at this point to discuss briefly the environmental impact of the car, in order to argue this. Cars are widely acknowledged as a main cause of many aspects of environmental degradation.[12]

They produce a range of pollutants, including carbon dioxide, which is the main gas producing climate change, nitrogen oxides (NOx) and volatile organic compounds (VOCs) which cause acid rain, a range of gases which produce urban air pollution such as carbon monoxide, VOCs, NOx, particulates, and others. This pollution causes system-wide environmental change on a grand scale such as climate change, leading to temperature and rainfall changes, sea-level rise, and so on, down to micro-level impacts such as a wide range of impacts on human, animal, and plant health. A point worth concluding here is that for most of these gases, motor transport is the only source which is, in the 'industrialised countries' at least, still increasing.

The second major class of environmental problems which a car economy is heavily involved in producing is resource depletion. Cars consume between 35 per cent of the oil in Japan and up to 63 per cent of the oil used in the US, simply in their use. Oil is also a major resource in asphalt and therefore road production. In the US, car production consumes 13 per cent of all the steel, 16 per cent of the aluminium, 69 per cent of the lead, 36 per cent of the iron, 36 per cent of the platinum, and 58 per cent of the rubber (both natural and synthetic) (Freund and Martin, 1993, pp. 17–19).

Finally, a car-based society has radically altered space. Urban space in particular has been systematically reconstructed to make allowance for the space required to move people about in cars. Cars take up huge amounts of space which could be used for other purposes. The highest figure is for Los Angeles, where two-thirds of all land space is devoted to car use – driving, parking (at shops, work, home, restaurants, and so on). For the US as a whole, about half of urban space is devoted to car use, while 10 per cent of available arable land is taken up by roads and parking places (ibid.). Many suggest that this has become a self-reproducing trend, as the reorganisation of towns and cities to make car-based mobility more possible has meant that increasingly a car has moved from being a luxury to a necessity (e.g. Gorz, 1980, pp. 69–77; Illich, 1974; Wolf, 1996).

The point to emphasise here is that the way that cars are usually discussed in academic and policy literature on global environmental change is with respect simply to one environmental problem, taken in isolation. So, regarding acid rain, the car is discussed with respect to nitrogen oxide emissions, and the solution is (claimed to be) catalytic converters.[13] Regarding global warming, the discussion moves to carbon dioxide, and the solution is portrayed as fuel efficiency and in the longer term a move to non-petroleum-based car engines. Regarding local air pollution, catalytic converters and lead-free petrol are the solutions. Regarding congestion, road pricing, more road-building, and perhaps some switching to public transport. The effect of focusing on single problems is to suggest that there are technical fixes, and some socio-technical fixes (public transport) which ultimately retrieve the car. But if we look at the impact of the car in its totality, the likelihood of success of such technical fixes declines significantly. However much we reduce NOx emissions, congestion remains; however much we introduce electric cars, these may simply become 'elsewhere-emitting vehicles'; however much we reduce the environmental impact of the materials used in car construction, they will still kill people on impact; and so on. Ultimately, producing a sustainable economy must involve a shift away from a car-dominated economy.

Cars and the IR of the environment

Despite what has been argued above, cars do not feature much in discussions of global environmental politics within International Relations (or in politics more widely). One reason for this is the focus within IR on particular environmental problems, abstracting from the broad interconnections between these problems. Another, as argued throughout this book, is the exclusion of questions about underlying causes of environmental change.

Where cars might appear is as particular objects of state or interstate regulation with respect to global warming or some other problem. But the silence is therefore more interesting given that in policy debates concerning, for example, global warming, it is widely recognised as being particularly problematic for states to regulate cars in order to reduce emissions.

So an argument that global environmental change goes to the heart of modern power structures and social processes becomes all the more persuasive in this case. The global politics of the car disrupts main-stream notions of the neutrality of the world's power structures with respect to environmental change. Cars show the implication of those structures in both producing environmental change, and hampering efforts to address that change. The politics of environmental change is therefore a politics of resistance.

The net effect of the combination of the fostering and emergence of a car culture, and active promotion of the car by states, has been a dra-matic shift from rail and public transport to the private car as the dom-inant means of transport. The main point to emphasise is that this development has been anything but inevitable or 'natural'. It has been produced both by states directly putting resources into road construc-tion and other subsidies to the expansion of the car, and participating (for its own reasons of legitimation) in the production of a car culture where (particularly masculine) twentieth-century identities are pro-duced through motifs such as speed, which the car (at least in its dominant construction) embodies.

This suggests that dealing with environmental change means much more than simply persuading governments to tweak policy instru-ments to achieve particular goals. The car is not something which people relate to in the rationalistic fashion presumed in the economics-driven world of policy analysis. It has been important to the continual reconstruction and performance of gender divisions, and the reproduc-tion of patriarchal power (through differential access to mobility, space, and so on). In any case, governments are themselves complicit in promoting a car economy, which has been useful for them both in relegitimising their rule – both directly, as promoting the car valorises dominant themes of modernity consistent with the values of the mod-ern state; and indirectly, as the car economy has stimulated economic growth at important periods, thereby helping to promote government legitimacy. Reducing car ownership and use therefore produces certain contradictions for the state – an environmental legitimation crisis, as analysed by Hay (1994), whereby in the need to intervene eco-nomically to maintain economic growth, the state comes into contra-diction with the need to intervene with respect to environmental change.

Challenging car culture

Political contestation over the car is not new. Far from being the unproblematic technology which all people until the early 1990s have viewed as the promoter of human liberty and welfare, it has often been seen in a damaging light both in environmental and social terms. Before the late 1980s/early 1990s, two principal periods can be identi-fied when car culture came under particular challenge. The first is at the beginning of the car's existence. Initially, there were sometimes severe restrictions to be overcome. Culturally, a classic expression of opposition to the car is often taken to be Toad of Toad Hall in Kenneth Grahame's *Wind in the Willows* (1908):

> it is the motor car that overturns innocent stability, the golden age; aboard a car Toad becomes 'the terror, the traffic-queller, before whom all must give way or be smitten into nothingness and ever-lasting night'.[14]
>
> (Overy, 1990, p. 73, quoting from Grahame, 1908)

Cars were widely seen as a nuisance. The British Prime Minister Asquith, called them in 1908 'A luxury that is apt to degenerate into a nuisance' (quoted in McShane, 1994, p. 113). They were associated with danger, noise, dirt, and threatened to disrupt established modes of urban life. But very quickly, resistance to the car foundered on the iden-tification of the car as a symbol of progress. In Wolfgang Sachs' words:

> What critics of the automobile saw themselves confronted with in the debates of the time could be called the executive syllogism of competition-driven progress: (a) technological development cannot be stopped; (b) escape is not an option, so Germany [or Britain, France, the US...] must take the lead; (c) therefore, we are called upon to support the automobile and its industry with all the means at the State's disposal ... The world market cast its long shadow over debates about the meaning of motorization on native streets.
>
> (Sachs, 1992a, p. 27)

The second period of challenge was during the wave of environmen-talism in the late 1960s/early 1970s. In this period, the dominant chal-lenge was to an extent in the UK, but even more so in the heart of car culture, the US.[15] The challenges were rooted to a great extent in the effects of the Highway Aid Act of 1956. Increasingly, cars lost their romance for many Americans, and had become part of the problem facing American society (Davies, 1975; Gordon, 1991, p. 14).[16]

According to Berman (1982), the Federal Highway Program which the 1956 Act produced was a development of Robert Moses' grand schemes for reconstructing New York. Berman describes in great detail the destruction of the Bronx by the Cross-Bronx Expressway constructed by Moses, and the resistance to that construction/destruction. Later resistance to expressway building in New York, and to the use of Highway Trust money for expressway construction in other US cities was more successful, however (Berman, 1982, p. 326; Davies, 1975).

This resistance to road-building and the further development of a car-dependent economy was to a great extent local in character, in recognition of the way that such expressway construction necessarily destroyed city neighbourhoods, and in many ways was designed deliberately to do so (e.g. Davies, 1975). But they also took on an environmental character in the more narrow understanding of that term, in terms of an emerging understanding of the car's role in resource consumption and pollution, prevalent themes in that wave of environmental concern. For Wernick, this context of resistance to the car's dominance on social/environmental grounds explains shifts in the way car culture was promoted through advertising (1991, p. 78), as discussed above.

Such resistance to car culture, both in the 1960s and 1970s and again in the early 1990s, mostly took the form of resistance to road-building schemes. This is partly because this is where the destructive aspects of car culture are most visible, and also appear preventable. The road is clearly the condition of possibility of car driving. But once built, preventing car use is clearly much more difficult.

I began this chapter with a discussion of public discourse surrounding recent road protests in the UK. The point there was to illustrate the context within which contemporary promoters of car culture have had to operate. But those road protests also tell us something about the nature of environmental politics. For mainstream approaches to global environmental politics, with their exclusive focus on inter-state processes, such phenomena are simply absent, irrelevant. But if cars are such an important component not only of the production of global environmental change, but also of the social, economic and political processes which make up state power and thus make inter-state politics possible, then such an absence simply serves to reinforce the way that the states system, capitalism, patriarchy, and so on, are naturalised and made unchallengeable.

One interesting feature of this is the trajectory of roads protest in the UK. Alongside direct-action protests against particular road-building schemes have been two complementary types of protest which have made the link between road-building, car culture, and broader questions of political forms. Critical Mass, which has a 'long tradition internationally' (Doherty, 1997, p. 10), has been a loose unorganised network of pro-cycling protests, where once a month cyclists in a number of the UK's towns and cities have met and cycled round the city centre at a Friday rush hour, clogging up traffic (further than it already clogs itself). The point is to show how particular transport modes, and through them, forms of mobility and subjectivity, dominate, and to challenge that domination.

Reclaim the Streets has similarly concerned itself with reclaiming urban space as a public space, in this case for parties. This loose organisation has organised parties in public streets, shutting them down to traffic, both as a form of protest, but also to show how public space has been transformed and destroyed by the car and the shift from the boulevard to the highway, as analysed by Berman. This makes the links between a politics of resistance to the car and to the state explicit, revealing the complicity of the state in reproducing the dominance of the car, as such parties are in themselves illegal.

Secondly, such resistance itself also helps to illustrate the main point made in this chapter. Resisters to road-building in the UK have often been made up of an odd alliance of NIMBYs (Not in My Backyard) wishing to protect their local sites of amenity, and radical Green activists operating more on NOPE principles (Not On Planet Earth) There are many instances of movement from the former to the latter type, as involvement in protesting develops in protestors a deeper sense of the social changes necessary to create sustainable societies.[17] The understanding of the deeper Green activists of the way in which road-building and car culture are embedded in a set of power structures is clearly evident in the way they understand their own actions. This is perhaps shown most fully by Seel in his analysis of the Pollok Free State protest camp against the M77 extension in Glasgow, which he conceptualises as 'embryonic counter-hegemonic resistance'.[18]

This can also be seen in the way that 'mainstream' society responds to such challenges. In addition to the broad cultural responses by the media and in advertising campaigns outlined at the beginning of this chapter, the political responses by the state to the protests support such an interpretation. On the one hand, 'The fact that they need six hundred security guards and police to chop down fifty trees is very significant. The state has to go to war on behalf of the car culture and a multi-national company' (Jake, protestor at Pollok protest camp, quoted in Seel, 1997, p. 122). On the other hand, the state in the UK has brought in legislation specifically designed to respond to and crack down on the new forms of protest involved in the roads protests, in sections of the Criminal Justice Act of 1994 which introduced new restrictions on 'trespassatory assemblies' and a new offence of 'aggravated trespass' (Doherty, 1997; McKay, 1996; Seel, 1997, p. 110).

Resistance to the car (and by extension, to road-building) thus has often been hampered by the car's association as one of the ultimate symbols of modernity and modernisation. To oppose it is thus to oppose modernity itself (Berman, 1982, p. 294). As Berman writes:

> The developers and devotees of the expressway world presented it as the only possible modern world: to oppose them was to oppose modernity itself, to fight history and progress, to be a Luddite, and escapist, afraid of life and adventure and change and growth.
>
> (1982, p. 313)

Or in the blunter words of Sam Turk, Ben Elton's fictional director of the thinly disguised 'Global Motors UK', '*Objecting* to roads! ... But that's crazy! What are they going to object to next? Food? Don't they want to get from A to B!' (1991, p. 49).

But shifts in the identification of the car with modernity are occurring, produced in part by resistance to the car itself, in part by increased recognition of the downsides of car culture, and in part by the emergence of other technologies which are perhaps displacing the car's symbolic purchase. In Krämer-Badoni's words, 'the car is in the process of losing the attribute of modernity' (Krämer-Badoni, 1994, p. 348). These shifts create possibilities for alternative modernities, not based on car culture.

Nevertheless, challenging the car involves much more than can be produced by technical or policy fixes. It challenges two fundamental aspects of the organisation of modern societies – space, and identity. The spatial organisation of societies has been fundamentally restructured around the car, particularly in the urban areas where the majority of people live (Wolf, 1996, pp. 152–5; McShane, 1994, ch. 10), in such a way as to reproduce dependence on the car, and make it difficult in many places to envisage moving to a system not dependent on the car without significant spatial reorganisation of cities. Secondly, cars have become central to modern identities, particularly masculinities, organised around notions of progress linked to speed. In Wernick's terms:

> the spread of cars rapidly transformed the entire ecology of life, creating massive, dependent road-systems and transformed cities; while at the individual level, it accelerated private and occupational mobility, altering our whole sense of time and space.
>
> (Wernick, 1991, p. 71)

Consequently, most people now tend to express a fundamental ambivalence when asked what they feel about the(ir) car. 'Cars were identified [in interviews] both as the most visible threat to the environment, and as an essential part of people's daily lives which could not be done without' (Macnaghten and Urry, 1998, p. 237). The implication of Macnaghten and Urry's account is that it is not primarily a question of resolving disputes between groups of people within society, but rather of dealing with this fundamental ambivalence which is present within individual people. The depth of social change necessary to move away from a car-dominated system should not therefore be underestimated, as it is not just a question of changing technologies, economic policies, and so on, but deeply-rooted individual and collective identities.

6
Fast Food, Consumer Culture and Ecology

Objecting to roads!... But that's crazy! What are they going to object to next? Food? (Elton, 1991, p. 49)

Today, almost every direct action is embedded in an extensive political matrix. No description is more misleading than 'single issue politics'. The people who started the squatters' estate agency in Brighton and those occupying the derelict land in London, today, want to change the whole world, not just part of it.

(Monbiot, 1996)

Introduction: McLibel

Alongside roads and car culture, many of the highest-profile protest campaigns in the UK in the early 1990s were over food, or more particularly, over meat. One of these concerned the live export of veal calves destined for overseas rearing where the conditions the animals were kept in were held by protestors to be barbaric. The central issue was therefore the question of humanity's ethical obligations with respect to other animals. Another was over bovine spongiform encephalopathy (BSE), or 'mad cow disease', where substantial numbers of British beef cattle were regarded as being infected with a disease which, it increasingly came to be believed, could be transmitted to humans and become Creuzfeld–Jakob disease. This conflict was over a broader range of issues, primarily threats to human health, but also government legitimacy, particularly concerning its credibility in making pronouncements concerning human health, animal welfare and the intensive-industrial methods of animal rearing, and the authority of science. It could also be interpreted as being about the policing of boundaries between humans and animals, the so-called 'species barrier'. Much of its political force derived from the permeability or otherwise of this 'barrier'. A third food campaign has been over genetically modified food. General food scares concerning potential health implications of genetic modification – the environmental implications of releases of genetically modified organisms into ecosystems have led to public campaigns, as well as, during 1998 and 1999, direct action against fields of genetically modified crops (the first trial for such criminal damage started in August 1998), and reactions from the companies involved, notably a very high-profile advertising campaign by Monsanto.

But along with that over genetic modification, perhaps the broadest conflict in terms of the coverage of issues was what became known as the McLibel case. In this libel case, easily the longest trial in British history, the whole range of operations of the world's largest food service corporation (Tansey and Worsley, 1995, p. 136), and largest owner of retail property in the world (Cummings, 1999, p. 16) was opened for scrutiny, from many different angles or points of view. Animal welfare, employment practices, tropical and temperate deforestation, packaging and waste, nutrition, advertising, form only a small part of the questions which came up for debate.[1]

McLibel started with the decision by McDonald's to issue writs in September 1990 against five activists from London Greenpeace, an anarchist collective (with no relation to Greenpeace International). The writs claimed that a leaflet distributed by the group (as well as by other groups) called 'What's Wrong with McDonald's' was libellous; that the claims in it about corporate malpractice by McDonald's were false and damaging to the company's business interests. Three of the activists apologised to the company, unable to envisage challenging the case. Two, Helen Steel and Dave Morris, decided they had nothing to lose and to fight it.

The trial closed finally in December 1996, after two-and-a-half years in court, preceded by three-and-a-half of preparation. The judge's verdict (he had ruled against a jury trial on grounds of the complexity of some of the evidence) took another six months to deliver. There were many things of interest in the trial. It has sparked a debate about the nature of British libel laws which uniquely require defendants to prove that their claims are true rather than plaintiffs to prove them false (one of the themes in Vidal, 1997; also Gorelick, 1997). The law is complicit in upholding corporate power behind the façade of legal equality, through enabling large corporations to take on legal personality in cases such as defamation, in combination with more obvious features of the system which the case made evident, notably the lack of availability of legal aid in libel cases, which resulted in Steel and Morris defending themselves.

It has also, perhaps ironically, helped to stimulate further resistance against McDonald's and the 'McWorld' more generally; generating an estimated 1.5 million leaflets distributed since the trial started (Vidal, 1995b; see also Beder, 1997, p. 69, who gives a figure of two million), pledges from thousands to continue distributing them whatever the verdict, events like the anti-birthday party at McDonald's headquarters on its 40th birthday (Bell and Valentine, 1997, p. 109), and the launch of the McSpotlight website in 1996, which focused on the trial and provided information and ideas for campaigners around the world, and was reputedly accessed seven million times in its first year (Vidal, 1997, p. 310). The very act of libelling those who objected to its practices became a focus of protest against McDonald's, exemplified by the name McLibel (alongside 'McMurder', McExploitation', and the other 'Mc' names given by protestors to aspects of McDonald's practices), and the existence of the McLibel Support Campaign. McDonald's actions, and Steel and Morris's refusal to apologise, as many others had had to in previous libel threats brought by the company (see Vidal, 1997, pp. 44–7, 124–6), were easily constructed as bullying tactics, of the large and powerful trying to silence their critics, and using the law as their ally. Steel and Morris made this explicit in their defence in the case; in Steel's words opening their case, 'We feel there is one word that can sum up what this case is about, and that word is "censorship". McDonald's are using the libel laws of this country to censor and silence their critics. ... This is a show trial against unwaged, unrepresented defendants' (quoted in Vidal, 1997, p. 100). The way that the case was readily constructed meant that the trial backfired on McDonald's (e.g. Bellos, 1995).

But what I want to emphasise is that McLibel, like the roads protests, shows that at issue in (global) environmental politics is a fundamental conflict over how the world should be organised. At root was a conflict over the sustainability and desirability of a globalised, corporate capitalist world order, exemplified by McDonald's. So McDonald's vs Steel and Morris exemplifies conflicts over sustainability; but to the protestors, these are fundamentally connected to questions of worker's rights, advertising practices, nutrition and health, and so on. As Brian Appleyard put it in *The Independent* (1994), 'It became clear what this trial is all about – the globalization of culture and belief systems.'

The depth of conflict over world-views in McLibel can be seen through the incommensurability of the arguments employed by Richard Rampton, McDonald's main lawyer, and the McDonald's witnesses, on the one hand, and the defendants on the other. Frequently, they could agree on 'facts', but what was at stake was the interpretation, the meaning of those facts. What conferred meaning on them were the fundamentally different world-views of the participants.

So, for example, McDonald's representatives expressed a continual incapacity throughout the trial to understand the motivation of the defendants. Mike Love (McDonald's UK Head of Communications), in response to questions from Steel and Morris about why he thought they and others protested against the company, stated that 'We can't predict why anybody would do anything to protest against McDonald's' (quoted in Carey, 1995). Vidal's book (1997) on the case enumerates many instances of simple, but deep, incomprehension between different players in the trial. Such incomprehension encompasses the large question of McDonald's failure to understand why the company might be a focus for protest, to a range of smaller ones. On the question of advertising targeted at children – one of the three issues where Steel and Morris won (in the original trial, they won one more on appeal in 1999) – what for Steel and Morris was exploitation of children, through 'pester power', to get parents into the restaurant, and increase sales, was simply normal advertising practice in a capitalist economy for McDonald's. On the questions of nutrition and of litter, the company argued that overall questions of nutrition depended on a whole diet, not just particular meals supplied by them, and that responsibility for that lay with the individual; similarly for litter, what happened to the packaging after it left the store was not their responsibility but that of the customer, although McDonald's has always had litter crews who cleaned up outside the store (how successfully was disputed during the trial). By contrast, Steel and Morris argued that McDonald's, playing such a large role in the diet of many people, and producing so much packaging that it would play a large role in the littering of areas near stores, had responsibilities in these areas which it was failing to live up to. The differences here are essentially differences between agency-oriented (where the responsibility lies with the person making particular decisions), and structural (where those who play a predominant part in structuring the decisions of others have responsibilities in that regard) ways of understanding the world.

These miscomprehensions do not seem best interpreted as disingenuousness – as deliberate attempts by either side to appear to misunderstand in order to score rhetorical points with the judge. For example, Love seems genuinely not to understand Steel and Morris, and the thousands of others like them. The trial was clearly about the 'truth'; that is, of course, the essence of trials in general, but of a libel trial in particular, where it is the veracity (as well as the damaging nature) of people's claims which are at stake. The point of the trial both for Mike Love and for Dave Morris was the 'truth'. ' "People are entitled to exercise their freedom of speech and to demonstrate peacefully within the law," says Mike Love. "But we believe that those taking part should look at the facts and be aware of the truth." ' But for Morris, McDonald's has forced its way into the public consciousness and yet is trying to suppress dissenting voices and alternative views of what it really represents. ... The truth is always worth defending' (quoted in Carey, 1995).

But the point is that the incommensurability in the exchange described above between the arguments of the defendants and Love shows that 'the truth' is discourse-dependent. What constitutes 'bad wages' is not something which can be decided outside of discourse. Discursive frames produce the criteria by which the 'badness' of wages are to be judged. So for Preston, these are clearly produced by market forces and minimal regulation by governments, while for Steel and Morris, alternative notions of the good life are invoked. Clearly then, this is ultimately a struggle over what the good life should be, as much as over the 'truth' or otherwise of particular allegations made by the 'What's Wrong with McDonald's' leaflet. McDonald's represents corporate capitalism, whose discursive frames produce a notion that 'bad wages' (or many of the other criteria) are simply those which are illegal.[2] For the defendants, bad wages are endemic under capitalism where corporations, aided by states, are able to exploit workers through low wages.

The conflict over McDonald's in the McLibel case was therefore simultaneously symbolic and material. It was symbolic in the sense that Steel and Morris clearly saw McDonald's as symbolic of a broader global corporate capitalist order, and as one of its biggest representatives (and particularly big in how it is directly present in many people's daily lives). But that symbolic struggle was simultaneously material, in that what was at issue were the practices of a large corporation (and by extension the broader system of which it is a representative) in relation to how it uses and transforms both people and the rest of the natural world.

Drawing on this argument about the nature of McLibel, this chapter argues that the practices exemplified by McDonald's are best interpreted as ecologically problematic, not solely in terms of particular instances

of environmental or social abuse, but as systematic and routine degra-
dations (Saurin, 1996). This is understood by anti-McDonald's campaigners such as Steel and Morris, for whom a focus on McDonald's
is essentially strategic, focusing on how to transform the agency of
those for whom a Big Mac is part of their normal lives. To do this, I
first look at the literature on 'McDonaldisation' produced primarily by
the work of George Ritzer (1996). Here, the practices of McDonald's are
interpreted as representative of, and in some senses driving, much
broader contemporary social trends. After discussing this, and critiques
of Ritzer's thesis, I look at two sets of themes embodied in the consumption practices exemplified by McDonald's which can be seen as
ecologically problematic: (intensive) meat consumption, and the acceleration inherent in the notion of 'fast' food. Finally, I revisit questions
of resistance in the light of the foregoing arguments.

McDonaldisation: McDonald's as modernity and modernisation

In a widely cited work, George Ritzer interprets McDonald's in terms of
Max Weber's theory of rationalisation (Ritzer, 1996). He suggests that
'McDonaldisation' is an appropriate term for dominant social trends.
He defines this as:

> the process by which the principles of the fast-food restaurant are
> coming to dominate more and more sectors of American society as
> well as the rest of the world.
>
> (ibid., p. 1)

Following Weber's analysis of rationalisation and bureaucracy, Ritzer
suggests that there are four major aspects of such a process: efficiency;
calculability (defined as quantification); predictability; and control/
replacement of humans by technology. The first of these involves
'choosing the optimum means to a given end' (ibid., p. 35), and
describes the process by which work is accelerated. Ritzer cites
Taylorism as the founding moment of twentieth-century concerns with
efficiency. For Ritzer, McDonald's managed to increase efficiency in
food-service by Taylorising food-service production (primarily involving the intensification of the division of labour, also involving some
use of new technologies), through simplifying the menu greatly to
make assembly-line production possible, and through putting the
customers to work in terms of queuing and clearing away the tables

(Ritzer, 1996, pp. 38–41). Ray Kroc, founder of the McDonald's empire, was widely regarded as obsessed with efficiency.[3] In Kroc's words, 'I put the hamburger on the assembly line' (quoted by Boyle, AP journalist, cited in Love, 1987, p. 211).

The second aspect of rationalisation concerns the increasing quantification of economic and other transactions; in this case, the increasingly detailed specifications concerning the size of a burger (1.6 ounces) and a bun, the amount of onions, the length of grilling time, the length of frying time for french fries, and so on. It also involves quantifying the time taken to serve a customer, so that success and quality becomes defined in terms of how many customers are served within three minutes of entering the restaurant (Ritzer, 1996, pp. 60–4).

The third concerns standardisation of the production process across all McDonald's outlets, so that customers know they will get exactly the same product wherever they buy their Big Mac. Ritzer quotes Leidner's book *Fast Food, Fast Talk: the Routinization of Everyday Life*: 'The heart of McDonald's success is its uniformity and predictability … [its] relentless standardization' (Leidner, 1993, quoted in Ritzer, 1996, p. 80). Predictability occurs in the replication of settings, in the scripted interaction between McDonald's workers and customers, in predictable employee behaviour, in predictable products (Ritzer, 1996, pp. 80–2). The predictability in the final category is as total as technically possible: hamburgers are regulated in their fat content to high degrees of standardisation; settings and frying times for french fries are standardised and mechanised; types of potatoes, methods of freezing, and so on, are all ruthlessly specified in the 600-page McDonald's operations manual.

Finally, increasing use of robots and technologies of surveillance enable both direct increases in efficiency, calculability and predictability, and also control of both workers and customers (ibid., ch. 6). This is the area where Ritzer suggests McDonald's have not fully rationalised their operations, but this is in part because he takes technology to mean machinery; with a broader notion of technology, the operations specified in the operations manual become instruments of employee control.

Ritzer then discusses the 'irrationality of rational systems'; that systems organised along such rationalised lines produce outcomes which are in practice irrational. More precisely, they are unreasonable in the sense that they are dehumanising. Ritzer suggests this irrationality has a number of features. He highlights some health and environmental consequences, focusing on questions of nutrition (p. 129), food poisoning (p. 129), the amount and (lack of) biodegradability of packaging producing large volumes of waste, the heavy use of paper having impacts on deforestation, and the use of particular materials such as polystyrene having particular impacts on landfill use and some specific environmental problems such as the consumption of CFCs which destroy the ozone layer (ibid., pp. 129–30).

He also argues it has a dehumanising effect on employees and customers, because of the minimal skill required by each, because of the 'assembly line eating' it produces, and because of the scripting of the relationship between customers and employees (ibid., pp. 130–6). Finally, he suggests the homogenisation of diet and eating throughout the world is dehumanising and irrational (ibid., pp. 136–7).

Ritzer's book has been widely cited. It has had entire books devoted to it (Alfino, Caputo and Wynyard, 1998). For example, simply in relation to food, a number of authors have discussed his thesis. For example, Bell and Valentine (1997, p. 6), Warde (1997, p. 17, 37) and Beardsworth and Keil (1997, pp. 120–1) all discuss Ritzer's thesis in terms of modernisation. Beardsworth and Keil emphasise the application of scientific management to food service, and Ritzer's appeal for 'slow food'. Warde suggests that Ritzer's account is a 'massification thesis' where tastes become more homogeneous as corporate capitalism develops. Warde suggests Ritzer's work can be taken as an example of Adorno's account of mass consumption.

But the more general notion that McDonald's represents many important features of modernisation is widespread, going beyond discussion of Ritzer's work. One widely discussed moment is the opening of the first McDonald's in Moscow, after the end of the Cold War, where the meaning of McDonald's is interpreted as signifying the arrival of modernity as consumer capitalism in the former Soviet Union (Bell and Valentine, 1997, p. 190; Smart, 1994, p. 175). Fiddes quotes a Russian journalist saying 'It's like the coming of civilization to Moscow' (1991, p. 66, quoting Reuter, 1990, p. 20).

The most predominant discussions of McDonald's as modernity, however, come in literatures on consumer culture, particularly with respect to debates about globalisation and homogenisation. Bell and Valentine (1997) begin their book with dialogue from the film *Pulp Fiction* in which the differences between McDonald's in the US and in Europe are discussed. McDonald's is taken as a paradigm case of globalisation. In Bell and Valentine's words, 'virtually everyone' uses McDonald's to 'think global' (1997, p. 190). Waters, in his overview of debates on globalisation, suggests that the saturation of US fast-food markets led McDonald's to develop a globalising strategy (1995, p. 70). Love (1987) makes a discussion of the globalisation of McDonald's operations the final chapter in his corporate history of the company. Finkelstein emphasises the importance of the uniformity of fast-food operations being global (1989, pp. 11–12, 93). Beardsworth and Keil suggest that the standardisation of operations is something which gives confidence to consumers in the face of 'increasing levels of glob-alization of culture and increasing levels of both social and geographical mobility' (1997, p. 170).

It remains debated, however, whether the global spread of fast-food operations, exemplified by McDonald's, is best interpreted as a homogenising trend. Bell and Valentine (1997, p. 190) suggest there is a prevalent homogenisation thesis, where globalisation is interpreted as 'McDonaldisation' or 'Coca-colonialism'. Ritzer's thesis can be taken as an example of this (1996). Warde (1997) argues that Ritzer gives a version of a 'massification thesis', similar to that developed by Frankfurt School writers such as Adorno, where 'differences of class, gender and nationality fade before the ubiquitous presence of McDonald's burgers and Coca-cola' (1997, p. 17). Fantasia's analy-sis (1995) of 'Fast Food in France' argues that it does represent a process of Americanisation of French culture. '"Fast food in France" has had less to do with food than it has with the cultural representations of Americanism embodied within it', he argues (1995, p. 229). Star (1991) also gives an account which relies on a version of a homogeni-sation thesis, through a focus on the exclusions created by rationalised production such as at McDonald's: 'McDonald's appears to be an ordinary, universal, ubiquitous restaurant chain. Unless you are: vege-tarian, on a salt-free diet, keep kosher, eat organic foods...or are aller-gic to onions' (Star, 1991, p. 37). Evidence for the homogenisation thesis could be drawn from Love's corporate history of McDonald's. When the company expanded its operations beyond the US, it initially tried to adapt the menu to indigenous tastes (for example in Germany, the Netherlands, Australia), but these attempts failed and it moved back to producing exactly the same menu as in the US. As it did this, however, it tried to produce a 'local' image, sensitive to charges of economic/cultural imperialism.[4] Therefore the marketing and store design but not the menu, were indigenised (Love, 1987, pp. 420–40, for Australia, see also Probyn, 1998, p. 160).[5]

But the dominant trend in this literature is to resist this homogenisa-tion narrative. Arguments that globalisation and McDonaldisation equal homogenisation are interpreted as denying the agency of consumers in constructing meaning through their consumption practices (Parker, 1998). Thus, McDonald's is held to have very different meanings in 'Moscow, Manchester or Michigan' (Smart, 1994; Bell and Valentine, 1997, p. 190). Perry (1995) similarly argues that such 'McTheorising' 'subsumes cultural meaning under social and/or economic relations' (as cited in Bell and Valentine, 1997, p. 11). Thus Bell and Valentine argue, while discussing the argument of Star mentioned above, that an equally persuasive interpretation of contemporary trends is to one towards hybridisation of consumption practices towards food (1997, p. 135). Probyn (1998) provides a slightly more nuanced version of this argument. She suggests that McDonald's produces homogenisation, in part by using notions of family and food to imply that since we all [*sic*] eat McDonald's, we are all one global family, as we eat together. But at the same time the company attempts to respond to concerns of diversity by 'tak[ing] up the trend towards hyphenating ethnic identities (e.g. Italian-Australian, Chinese-Canadian) and gives us an identity as McDonald's-world-Australian or McDonald's-fill in the blank' (Probyn, 1998, p. 155). However, whether the homogenisation thesis is accepted or not, it remains the case that these interpretations of McDonald's and the social trends it exemplifies are ones which are regarded by the above writers as integral to modernity and modernisation.

Martin Parker's discussion (1998) of Ritzer's work is particularly interesting for the present purposes. Parker argues that Ritzer essentially offers a conservative elitist critique of mass consumer culture. This suggests that a mass culture inevitably erodes the 'superior' culture of elites and that the critique is Romantic and nostalgic. Parker traces this critique back to 'Arnold, Leavis, Eliot, Nietzsche' and contrasts it with a 'left' critique (1998, pp. 2–5) which is nevertheless similar in many respects, but focuses on how a mass culture is constructed to preserve the power of capitalist elites, an argument coming primarily from the Frankfurt School (as noted above, this latter position is how Warde [1997] interprets Ritzer). Parker contrasts both of these critiques for mass consumer culture with the more affirmative stance of what cultural studies has become. The opposition is constructed as one of structure versus agency, with critics of mass consumption being regarded as overly structuralist, while the 'popular culture' writers focus on how consumer items are appropriated by their consumers and can become sites of resistance (see also Miles, 1998, Wood, 1998, and Wynyard, 1998, for similar arguments).

While Parker wants to retain a notion of critique, and be able to condemn McDonald's (or McDonaldisation) for its dehumanising labour process, ecological consequences, and so on, he makes this by and large a matter of social ontology; the critique pertains to our own social position and perspective. The critique then becomes simply what McDonald's means for us, and we are left without the possibility of judging that critique. Parker concedes the lack of a firm basis for judging McDonald's (1998, p. 15), as does Alfino, who writes 'Postmoderns are entitled to *feel* all the moral concern they want, but without a normative theory (and a prior commitment to some metaphysic of morals) they have no *rational* basis for adjudicating Ronald McDonald's crimes against humanity' (Alfino, 1998, p. 186).

The danger here is that a false opposition is set up; any judgemental discussion of particular consumption practices becomes charged with authoritarianism and elitism. But this argument is dangerously close to neoliberal economic arguments about consumer sovereignty and rational agency (as exemplified in relation to Ritzer by Taylor, Smith and Lyon (1998). A rejection of Marxist notions of false consciousness or needs which underpins Frankfurt School arguments about mass consumption should not lead, either implicitly or explicitly, to an argument that mass consumption is unproblematic and can in some versions of the argument become a mode of liberation. It is still possible to argue that particular consumption practices and the identities constructed around them are produced through strategies of power, rather than simply existing as individually decided preferences, let alone as practices of resistance. As Kellner states:

> Ritzer's critics sometimes offer apologetics and celebration of the mass culture he criticizes thereby uncritically replicating a position increasingly widespread in cultural studies that puts all the weight of praxis and production of meaning on the side of the subject, thus effectively erasing the problematics of domination, manipulation, and oppression from critical social theory.
>
> (Kellner, 1998, pp. viii–ix)

Ecological critiques of mass consumption are interesting and useful here. While it is certainly the case that some aspects of ecological critiques have resonance with romantic anti-modernism, they undermine, if taken seriously, the notion that consumption is simply about an endless play of signification. The semiotic aspects of consumption, which most contemporary sociology of consumption focuses on, seems to me inadequate here (e.g. Lury, 1996; Slater, 1993; Featherstone, 1991; Bell and Valentine, 1997). Ecology itself becomes simply reduced to a meaning which particular consumer items, for example, the Big

Mac, have. The ecologist identifies the Big Mac as a product symbolic of the high-intensity consumption which is undermining the sustainability of the planet's ecosystems. As shown above in relation to the McLibel case, the focus on McDonald's is clearly symbolic; the company stands as a symbol for all which Steel, Morris, and many others, regard as a set of social forces to be resisted. But much of that resistance is based on a notion that the practices exemplified are unsustainable; that is to say, impossible in the long term. Unsustainability implies that a practice is impossible in the long run. It seems reasonable to suppose that a condition of possibility of placing any other sort of meaning on a Big Mac is that such an object can be presumed to exist, along with people to eat it.

An ecological critique of this sort challenges the predominant tendency within studies of consumption. It suggests that the material-ecological basis of consumption practices is something which necessarily places limits on the symbolic possibilities of objects of consumption. In the language of ecological economists:

> Every economic phenomenon ... can be described as a flow of material and energy which begins in the environment, passes through the factory, house, city, humanised territory ... and returns, sooner or later, to the environment.[6]
>
> (Nebbia, 1990, p. 80, cited in Hayward, 1994, p. 109)

At the same time, much ecological discussion of consumption is overly technical, and abstracts from the symbolic-cultural meanings underpinning consumption practices. Many treat consumption by and large in terms similar to the way it is treated by neoclassical economists – as the total amount of goods and services 'consumed', measured in money terms (e.g. Lintott, 1998; Redclift, 1996) – and thus miss the specifics of particular consumption practices in both semiotic and material terms. Redclift argues that both semiotic and material analyses of consumption are necessary (1996, pp. 4–6) although he concedes that his book does not engage in the former. He suggests that Lash (1990) is one of the few to have emphasised both (Redclift, 1996, p. 6). Regarding food, Goodman and Redclift (1991) attempt such an integrated analysis. While this ecological critique of consumption, concerned with both the symbolic and material, could take a number of angles, I will focus on two: meat and speed. Such themes feature as two of the oppositions made by Warren Belasco in his analysis of countercultural food politics – vegetable vs animal, and slow vs fast (Belasco, 1989, pp. 50–61).

Meat

It is perhaps banal to point out that McDonald's reproduces a meat-centred diet. But my argument is that pointing this out serves to emphasise the material, embodied, nature of consumption. Whatever the symbolic constructions surrounding McDonald's and a Big Mac, it remains the case that such consumption necessarily reproduces an intensive system of production and consumption of a narrow range of foodstuffs, centring on beef production. Consumption of meat itself is drenched with cultural meaning. The meanings of meat help to show how the heavily meat-intensive diets, which at least the affluent industrialised countries of the world have been able to sustain, are intertwined with the reproduction of various forms of social power. I will emphasise three themes here.

Masculinity, power and domination

The first is the association between a meat-eating culture and a patriarchal one. The classic analysis of this is Carol Adams' *The Sexual Politics of Meat* (1990).[7] Adams is not so much interested in the way in which meat consumption is distributed between men and women (as well as between [male] adults and children), which is a common assertion; but that meat is the food which men arrogate for themselves particularly (Adams, 1990, pp. 28–9; for works which emphasise this, see Kerr and Charles, 1986; Fiddes, 1991, pp. 158–60). She is more interested in the ways that meat-eating and patriarchy reproduce and reinforce each other. Her claim is more than simply that meat is symbolically produced as masculine food, a commonplace assumption (Adams, 1990, pp. 26–8, 32–4; also Fiddes, 1991); but that it helps to reproduce a form of masculinity which is patriarchal; a masculinity as dominance.

Her main focus is on the intertwined nature of 'absent referents' involved in meat-eating and in patriarchal domination. 'Absent referent' is used to mean that which must be made absent in order to make its domination/destruction possible. Thus, 'through butchery, animals become absent referents. Animals in name and body are made absent *as animals* for meat to exist' (Adams, 1990, p. 40). She suggests there are three forms of absent referent here: the literal absence of a live animal once it is killed; a definitional absence, where the name of the animal is eradicated (cows become beef, pigs become pork); and metaphorical, where for example the phrase 'I felt like a piece of meat' helps to metaphorically make actual pieces of meat, as dead animals, absent (pp. 40–2). But she then suggests that women and animals are overlapping absent referents, reproducing the social power which enables others to make them absent:

> sexual violence and meat eating, which appear to be discrete forms of violence, find a point of intersection in the absent referent. Cultural images of sexual violence, and actual sexual violence, often rely on our knowledge of how animals are butchered and eaten.
>
> (Adams, 1990, p. 43)

She then goes on to analyse parallel cycles of objectification, fragmentation and consumption in relation to women and animals, where first their subjectivity is denied, and they become objects for the use of men and humans in general. This objectification enables them to be fragmented, which in both cases she takes literally: butchery, or the 'disassembly line' for animals; and sexual violence, rape, murder, for women (pp. 49–61). Fragmentation then enables consumption (pp. 47–8).

Adams also makes the connection between patriarchy and meat-eating through connections between the movements opposing each of them: feminism and vegetarianism. She shows, primarily through literary analysis of works of prominent vegetarians and feminists, how feminist writers from the seventeenth century onwards have connected meat-eating with patriarchy and war, and vegetarianism, with peaceful, more equal societies: for example, in discussing the works of Mary Shelley (the author of *Frankenstein*), Charlotte Perkins Gilman (*Herland*), and Isabel Colegate (*The Shooting Party*). Thus, in her conclusion, she argues, that 'Meat eating is an integral part of male dominance' (p. 167) and 'Meat eating is the re-inscription of male power at every meal' (p. 187). As a result of this, 'vegetarianism acts as a sign of dis-ease with patriarchal culture' (Adams, 1990, p. 167).

The salient connection is perhaps that particularly modern ideologies and practices of domination, are constructed and understood as masculinist. Thus, as pointed out in Chapters 3 and 4 above, the domination of nature by humans, as a particularly modern phenomenon conventionally understood as being developed in science by philosophers such as Francis Bacon, was simultaneously a patriarchal project to control women's bodies and reorganise male power over women (Merchant, 1980; Shiva, 1988; Plumwood, 1993). Chapter 4 showed how sea defences can be interpreted as a particularly visible, high-profile, expression of such domination of nature. But meat-eating is also a prevalent expression of such domination. Although more mundane and everyday, it expresses human power over the rest of nature particularly clearly (Fiddes, 1991).

Modernity, progress and affluence

Clearly connected to the question of domination, the second theme underpinning meat-eating concerns the way that meat, like cars (as emphasised in the previous chapter), has become a symbol of modernity and affluence. Meat-eating culture has generally presented itself as representing such progress. Histories of diet generally assume or explicitly argue that there is a 'natural' tendency to consume more meat as societies get richer. Fiddes (1991, pp. 56–7) cites *The Hamburger Book* (Perl, 1974), as arguing that meat-eating was the mark of the emergence of civilization – the more civilised humans became, the more meat they consumed, and the more complexity became involved in the preparation of that meat. Adams also notes how this discourse of civilization = meat-eating often has racist connotations, being used for example to justify imperialism, and to explain the 'backward' nature of 'primitive' peoples. Cannibalism was often interpreted in terms of the lack of meat in the diet leading people to turn to anthropophagy (Adams, 1990, pp. 29–32).

The association of meat with modernity has produced a global politics in which increases in meat-consumption have been taken as indicators of modernisation in developing countries, and produced a North–South politics in which refusal of meat by westerners may be taken by people in the South as an insult. Such a dynamic is stimulated by global disparities in meat-consumption, paralleling global disparities in wealth and income. 'The developed world consumes roughly two-thirds of world meat production whereas the developing world with three-quarters of the world's population only consumes one-third of total meat production' (Williams, forthcoming, p. 6). Given meat's cultural position as a superior, 'modern', 'civilised' foodstuff, a cultural North–South politics of resentment is unsurprising, especially given that significant portions of developing-country vegetable exports (primarily foods such as soya and groundnuts) are exported to feed animals in the West.

Fast food may be seen as the ultimate in modernised meat-eating. This is where food service was initially industrialised and rationalised – recall Kroc's statement that 'I put the hamburger on the assembly line' quoted above. Combined with America's representation as the most modern country, the hamburger, particularly McDonald's, has become a potent symbol of modernisation as Americanisation (Fiddes, 1991, pp. 66–7).

Beardsworth and Keil (1992) illustrate a paradox of the relationship between meat-eating and modernity, however. This is that the 'vegetarian option', as they call it, is in some senses itself a product of the nutritional variety produced by affluence. There is a sense in which this argument presumes a myth of original meat-eating by humans, a myth which is heavily debated. More importantly, however, it remains the case that meat remains a symbol of affluence.

Resource intensity

The prosperity in industrialised countries which makes extensive meat-eating possible facilitates a substantially more intensive form of food production than can be sustained in non-industrialised societies. And it is commonplace to observe that meat production is significantly more resource-intensive than vegetable food-production. Adams notes that Plato argued that because meat production required large amounts of pasture land, it would lead to wars as neighbouring states competed over land for that pasture (1990, p. 115–16). She says that some feminist writers, such as Gilman, in her utopian novel *Herland*, picked up on Plato's argument as a reason for making her utopia vegetarian. Analyses of vegetarianism suggest that understandings of the ecology of meat production and its over-consumption of land is one factor which leads some to become vegetarian, although it is not as common a reason as health or animal welfare/rights beliefs (Beardsworth and Keil, 1992).

The ecological aspect of meat-consumption has two aspects. One is simply that meat-consumption requires much more throughput of resources than does a diet not involving meat. It is commonly observed that it takes 16 lb of grain to produce 1 lb of beef (Moore Lappé, 1982, p. 69). In the UK, 80 per cent of agricultural land is devoted directly or indirectly to meat and dairy production (Spencer, 1993, p. 330). Seager is worth quoting at length on such resource use:

> between 1960 and 1985, 40 per cent of all Central American rainforests were cleared to create pasture for beef cattle. ... Cattle ranching is responsible for an estimated 85 per cent of topsoil erosion in the US, and similar devastation in Australia and Canada. Within the US, half of all water consumed is used to grow crops that are fed to livestock; meat production requires at least 10 times more water than grain production. More than 50 per cent of water pollution in the US can be linked to wastes from the livestock industry, including manure, eroded soil and synthetic pesticides. ... Meat

production places enormous demands on energy: the 500 calories
of food energy from one pound of steak requires 20,000 calories
of fossil fuel.

(Seager, 1993, p. 211, citing Adams, 1991, and Kirchoff, 1991)

In addition, meat production, especially on an industrial scale,
intensifies use of other resources in agriculture. Cronon's classic
environmental history of Chicago (1991, pp. 206–59) gives us clues
here. In order to develop the mid-West agriculturally, to satisfy growing
demand for meat in the Eastern US cities, its whole ecology was
reorganised. Following the mass slaughter of the bison on the prairies,
and in part causing this slaughter, the plains were given over largely
to cattle, and indirectly through grain production, to pigs. This
involved a reorganisation of the ecology of the plains in accordance
with market logic, entailing a great intensification of production and
rationalisation of space. Refrigeration on railway carriages was developed in order to be able to ship meat (rather than live animals)
and thus accelerate the quantity of meat transported. The first mechanised forms of production developed later in Chicago and then
by Ford, were introduced in Cincinnati, or 'Porkopolis'. Known as the
'disassembly line' this innovation greatly increased the efficiency of
pork and beef production. Cronon suggests that one of the deepest
ecological consequences was that whereas the connections between
meat-eating and the animals and ecosystems this depended on were
previously visible (because local), the industrialisation of meat packing
made such connections obscure. Echoing Adams' notion of 'absent
referents', Cronon suggests this development produced 'unremem-
bered deaths' (pp. 247–59).

It is not then perhaps an accident that the world's largest food
service organisation, and a means of organising food consumption, are
centred on meat, resonant as both are with prevalent themes of modernity and domination. However, reproducing and expanding such
consumption practices have necessary ecological consequences which
make them problematic. The capacity of the world's soils to continually
expand their productivity to feed such resource-intensive modes of consumption is doubtful at best, and has consequences also for the distribution of food among the world's peoples. As Spencer puts it, 'it is
profoundly ironic that the human need [*sic*] to prove our dominance is
the driving force which exhausts the environment', (1993, p. 343). This
is resonant of Horkheimer and Adorno, for whom 'the fully enlight-
ened earth radiates disaster triumphant' (1979, in Hayward, 1994, p. 8).

The speed of fast food

The ecological consequences of meat production and consumption have been greatly intensified by the modernisation of the food industry, as exemplified by the 'fast' in fast food. As noted in the previous chapter, speed/acceleration are often taken as primary defining features of modernity – the continuous acceleration of life, and replacement of older modes of consumption with newer, faster ones – 'all that is solid melts into air', to repeat the famous phrase. Such industrialisation and acceleration has produced consumption practices which are steadily more intensive – people's intake of food in the West has increased throughout the 20th century, made possible by ever more 'efficient' means of food delivery.

Fast-food restaurants represent one important facet of this increased efficiency (along with supermarkets, refrigeration, and so on). It was the essence of McDonald's distinctiveness from its inception. Discussing the original restaurant in Pasadena, later moved to San Bernadino, run by the McDonald brothers, Love writes:

> Now they decided to make speed the essence of their business. 'Our whole concept was based on speed, lower prices, and volume', says [Dick] McDonald. ... 'Customers weren't demanding it, but our intuition told us that they would like speed. Everything was moving faster'.
>
> (Love, 1987, p. 14)

The McDonald brothers thus called their assembly line system the 'Speedy Service System'.

These intensified consequences in part relate to increased meat consumption, and increasingly to transformed conditions of production of meat. But they also relate to other aspects of environmental change – increased use of non-reusable packaging, increased use of energy as more food is refrigerated and transported over greater distances, increased monoculture in agriculture, coevolution with automobility and its consequences. The emergence of fast food restaurants has therefore set in train a dynamic which intensifies the ecological impacts of food consumption.

Fast-food restaurants are major consumers of the raw materials making up most of the products they sell. McDonald's alone is the biggest consumer of meat and potatoes in many of the countries it operates in, in the US consuming 600 million pounds of beef annually, and

7.5 per cent of the US's total potato food crop (Love, 1987, p. 3). Love's eulogistic corporate history is revealing in this regard; presented in terms of McDonald's huge commercial success, Love shows how some of the deepest consequences of that success are the 'revolutionary changes in meat and potato processing' (ibid.):

> In their search for improvements, McDonald's operations specialists moved back down the food and supply equipment chain. They changed the way farmers grow potatoes and the way companies process them. They introduced new methods to the nation's dairies. They altered the way ranchers raised beef and the way the meat industry makes the final product.... no one has had more impact than McDonald's in modernizing food processing and distribution in the last three decades.
>
> (Love, 1987, p. 119; also Cummings, 1999)

It is not difficult to envisage that such transformations are simultaneously ecological transformations. For example, concerning potatoes, McDonald's specified the use of the Idaho Russet as the ideal potato for making a McDonald's french fry. McDonald's therefore rigorously imposed use of this potato on its suppliers, producing increased monoculture in American (and later elsewhere) potato farming (Ritzer, 1996, p. 13). They also transformed the way potatoes were stored, electrifying this part of the process in order to produce a standardised product quality, and later changing over to frozen potatoes in order to be able to use the Idaho Russet all year round (the variety was unable to stand the summer heat) (Love, 1987, pp. 119–23, 330–5). Similar shifts occurred in beef production and processing, from reusable to disposable packaging, and later on in chicken production. Love suggests that as early as 1962 McDonald's had sufficient market power to enforce these changes on suppliers (ibid., p. 123). Others also suggest that large institutional buyers such as McDonald's hold great power over farmers and processors (e.g. Tansey and Worsley, 1995, p. 141).

Such organisation of consumption requires the intensification of agriculture, leading to what in animal farming is now usually referred to as factory farming. Ritzer (1996, pp. 112–14) discusses this as an aspect of McDonaldisation in relation to its 'control' dimensions. Such farming necessarily increases the strains on the animals that live on such farms, through feeding and breeding to increase bodyweight greatly, use of growth hormones to accelerate growth, producing bodyweights such that animals' legs are routinely in great pain just supporting them, intensive use of antibiotics, deprivation of space, light, and so on (see e.g. Singer, 1976; or Spencer, 1993, pp. 322–9, for a discussion of this). Fast food produces a dynamic whereby such conditions of production are necessary, since the volume of production could not be sustained without it.

One of the more controversial aspects of the McLibel case concerned the claim that McDonald's was involved in promoting deforestation. The claim was that land was being cleared to make way for beef farming for hamburger meat. This has been a longstanding argument of critics of fast-food restaurants. Again, the logic can be interpreted systemically; that the intensification of production and consumption represented by the fast-food industry necessitates increases in the land devoted to beef farming, with inevitable incursions into land which was previously forested.[8] The 'What's Wrong with McDonald's' leaflet made such claims about McDonald's; similar claims have been made widely about the fast food industry more generally (e.g. Fiddes, 1991, pp. 212–13; Transnationals Information Centre, 1987, p. 18).

Fiddes (1991, pp. 66–7, 232) suggests that the hamburger as an industrialised, highly processed form of meat is itself a response to increasing concerns about the ethics and ecology of meat consumption produced by urbanisation. Hidden in a bun, and transformed from any obvious connection to the body of the animal(s) which produced it, the meat in a hamburger can be readily disconnected from its animal origins.

> Like so much industrial production, the mass-produced hamburger effectively divorces consumption from its ecological context. Fast-flesh emporia entice the consumer with sanitised gratification; here everybody smiles, while health, welfare, and environmental implications are banished to another less seductive world.[9]
>
> (Fiddes, 1991, pp. 66–7)

Competitive wars between fast-food restaurants have led to phenomenon such as 'supersizing', where the restaurants have produced progressively larger-sized portions to gain competitive edge over their rivals. McDonald's introduction in the late 1960s/early 1970s of the Big Mac and the large portion of fries was to help McDonald's compete with Burger King and Wendy's, but in the early 1990s another round of increasing sizes gripped the fast-food industry (Wroe, 1996), producing moral panics over the health threats (Bell and Valentine, 1997, p. 135), but also intensifying further the ecological consequences of food production and consumption.

Conclusions: resisting 'McDonaldisation'

From the point of view of this discussion, Ritzer's argument is promising but ultimately frustrating. He provides an argument that the fast-food restaurant, with McDonald's as its paradigmatic case, is exemplary of processes of rationalisation, and that the organisational changes instituted initially by the McDonald brothers and developed by Ray Kroc, who turned McDonald's into an empire, have then been broadened out into many other economic sectors and areas of life (he discusses education, health and the workplace at length, and many other aspects of life). He also shows many of the downsides of such developments, although he also suggests they have advantages also (reminiscent of Berman's (1982) arguments concerning the contradictory nature of modernity and modernisation). But having done this, his discussion of whether the 'iron cage' of McDonaldisation can be escaped is very frustrating. He suggests that the process is ultimately unchallengeable, without really arguing it in the relevant chapter (entitled 'The Iron Cage of McDonaldization'),[10] or considering resistance to McDonald's or McDonaldisation at all.

The process Ritzer outlines apparently occurs almost outside human agency. 'McDonaldization has an inexorable quality, multiplying and extending itself continuously', he writes (ibid., p. 161). Earlier, he suggests its inexorable quality by suggesting that even when McDonald's and the principles of organisation it developed are gone, rationalisation will go on:

> When McDonald's has, like its predecessors, receded in importance or even passed from the scene, it will be remembered as yet another precursor to what is likely to be a still more rational world.
>
> (Ibid., p. 160)

Thus, 'McDonaldized systems will remain powerful until the nature of society has changed so much that they can no longer adapt to it' (ibid.). This is both a frustratingly depoliticised interpretation, and a contradictory one; elsewhere, McDonaldised systems are at the forefront of social change; now they appear as subject to the whims of social change. But politically, it again suggests that these processes are not produced by human agency. Ritzer is evidently aware of this criticism. He does claim that 'Lest I be accused of anthropomorphizing and reifying McDonaldization, it is actually people and their agencies that push the process' (ibid., p. 229). But this is in an endnote; nowhere in the text do the 'people and their agencies that push the process' appear, and then by definition nowhere do the people and their agencies who resist the process and argue for alternatives appear. Since resistance is impossible, and Ritzer is not entirely clear that it is desirable, his argument is that all that can be done is to take steps to 'humanize a McDonaldized society' (ibid., pp. xx–xxi). As Rinehart (1998) argues, concerning this conclusion, Ritzer thus views all agency in excessively individualistic terms, neglecting to consider possibilities of collective action to resist rationalisation.

Rather, McLibel and the argument above suggest that resistance to McDonald's and McDonaldisation is possible. On the one hand, such resistance has clearly had effects in transforming some of the practices of institutions like McDonald's. Lawson (1992, pp. 85–6) shows how McDonald's has often taken the lead in responding to criticisms, for example by reducing paper-use by switching to polystyrene packaging in 1976, and then back to paper in 1990 because of ozone depletion and landfill concerns, by increasing recycling, and by improved energy management, involving a 45 per cent reduction in energy-use at a test restaurant. At the same time, there is much 'Greenwash' and PR spin put on such changes which overestimate their impact (Lawson, 1992, p. 84; Beder, 1997, p. 171). McDonald's have performed such Greenwash by 'forming a partnership with the Environmental Defense Fund' (Beder, 1997, p. 132; Rowell, 1996, p. 109; Karliner, 1997, p. 192). Such PR efforts are at times helped by academic apologists for their operations, often in management or business studies. Lawson claims, for example, with obvious ideological effect, that 'the contribution which the fast-food industry can make towards mitigating the effects of major environmental problems, such as the greenhouse effect, acid rain and the erosion of the ozone layer is probably minimal' (1992, p. 184). As Belasco suggests in a different but related context, 'what was significant was that these campaigns [to persuade people of companies' green credentials] had to be waged at all. A more secure establishment would have had to say nothing … persuasion (or force) is needed only when authority breaks down' (1989, p. 130).

But at the same time, that such authority breaks down is insufficient as evidence that more far-reaching changes will (or even can) occur. The analysis above suggests that the effects of resistance designed to promote reform of particular practices by McDonald's and similar organisations is always likely to be only ameliorative. For it is the rationalisation which is at the heart of such operations which is at issue. Resistance is better thought of as building alternative forms of food provision to the industrialised, rationalised system of provision exemplified by McDonald's.

The critiques of McDonald's and McDonaldisation by Steel and Morris and their allies hark back to the countercultural critique discussed extensively by Belasco (1989). The counterculture in the US challenged the dominant modernist food industry over issues of ecology, centralization, patriarchy, health and speed. They made food a centrepiece of their broad social critique. Belasco recounts how these critiques became either challenged by hegemonic culture and/or coopted by it. As a consequence, the whole countercultural critique of prevailing social organisation became transformed into an individualist consumerist set of concerns over health (narrowed to particular questions such as cholesterol or sodium) and 'nature' (captured by big business's rendering the term meaningless). The anti-McDonald's campaigns revitalise the earlier food politics. At the same time, Belasco shows (1989, ch. 4) how the alternative food culture created an alternative food economy and infrastructure. Like McDonaldised systems, this is simultaneously a social and ecological transformation, but unlike the former, it is one premised on principles consistent with ecological sustainability. I turn now in the final chapter to examine how we might conceptualise this political resistance, and what forms of political action and community such resistance might lead to.

7
Conclusion: Globalisation, Governance and Resistance

In the previous four chapters I have tried to establish primarily that the power structures of global politics, as outlined in Chapter 3, systemically generate global environmental change. But Lenin's question, 'What is to be done?', remains. For mainstream writers on environmental change in IR, the answer to this question is clear. State elites should build stronger international institutions to address such change more effectively than they have to date. Environmentalists should persuade and pressure those elites to build such institutions. Occasionally, perhaps, environmentalists can participate directly in fulfilling governance functions themselves, but this is relatively marginal to the central feature of global environmental politics, which is interstate management. My intention has been to destabilise the assumptions on which this normative vision of international environmental politics rests. But there are questions of political action still to be addressed. Two in particular are most relevant here. Firstly, if capitalist, statist, scientistic, patriarchal structures are intrinsically unsustainable, what forms of political and social structures are consistent with principles of sustainability (defined in terms of the argument of Chapter 3 as those which do not require accumulation, and are not based on modes of domination)? And secondly, what forms of political action might help to move societies from 'here' to 'there' (or, as Mary Mellor puts it [1995], emphasising Green localist concerns) from 'there' to 'here'?

Global civil society and global environmental governance

A useful way into these debates, as mentioned in Chapters 2 and 3, is through the emerging literature on 'global civil society' and 'global environmental governance'.[1] As mentioned in Chapter 2, there is a growing set of writers within IR who are moving away from strict notions of governance as a network of interstate systems, international regimes. This is even the case for mainstream liberal writers such as Oran Young (1997a). But in Young's work, there is a clear tension between this development and the disciplinary commitments of orthodox IR, which is necessarily state-centric.

Wapner (1996) gives an account of three perspectives regarding the state system – statist, suprastatist and substatist (see also Hurrell, 1994). For Wapner, all three have significant commonality in focusing on the states system, either as the locus of effective responses to global environmental change, or as the core problem for global environmental politics which needs to be transcended. Wapner then contrasts this to what he calls 'world civic politics', which consists of the practices of transnational environmental groups (he discusses Greenpeace, the World Wide Fund for Nature and Friends of the Earth), which politicise global civil society in various ways. For Wapner, this is a way of circumventing the state-centrism of IR and offering an account of global environmental politics not based on regimes or other interstate processes.

Wapner's focus is very useful in debunking notions that politics only takes place in or between states. But he makes a number of moves which I want to take issue with here. Firstly, I am not convinced by his argument that substatists are committed to the states system in the way he suggests. Clearly, he is right to suggest that they focus on the states system as a generator of global environmental change and as a constraint to achieving sustainability. But his suggestion is stronger than this; effectively, he argues that such writers are committed to a model of politics which is state-like. Decentralisation of power, as Wapner reads the substatists, is simply a matter of recreating existing political institutions, sovereign states, at much more local, 'human scale' levels. But this is a misreading. Such Green decentralists do base much of their arguments on questions of scale. But they are also clear that such decentralisation for ecological purposes involves creating fundamentally different political institutions. That is clear in the way that many such writers are explicitly opposed to institutions and practices of sovereignty; as Helleiner (1996) points out, this has always been an intended implication of the slogan 'Think Globally, Act Locally' (see below). It is also clear that such decentralisation arises from Green concerns with hierarchy and domination. So the state is, for the substatist position, not simply about the scale of political institutions, but also their form.

Secondly, Wapner ducks a question which I have argued should be a central component of any account of global environmental politics; that of the causes of global environmental change. In his accounts of supra- and substatism, such concerns are clearly mentioned. However, when he goes on to discuss world civic politics, such questions suddenly disappear from view. But it is not clear how Wapner moves from a discussion of how the tragedy of the commons, or alternatively hierarchy and domination, or 'bigness', systemically generate environmental change, to the focus on politics in the way he does. Isn't some notion of the appropriate forms and scales of political institution necessary for the notion of world civic politics to be persuasive? What Wapner's analysis lacks here is a connection to some substantive outcomes. How does world civic politics provide a model of a system of governance which can in principle generate sustainability? Wapner's answer to this, given in another context, could be that there is no single answer. World civic politics is just one among many mechanisms which can be developed to help produce sustainable futures. But there is still for me a contradiction here since some of the other mechanisms are systemically anti-ecological.

Lipschutz (1997; also Lipschutz and Mayer, 1996) has a very similar conception of an emerging pattern of global environmental governance. He outlines a common distinction between government and governance, where as opposed to the reliance on enforcement through law and force, as is typical of government, 'governance is … a system of rule that is as dependent on intersubjective meanings as on formally sanctioned constitutions … of regulatory mechanisms in a sphere of activity which function effectively even though they are not endowed with formal authority' (Lipschutz, 1997, p. 96, quoting Rosenau, 1992, pp. 4–5). Patterns of global governance are therefore a mix of interstate regimes (as focused on by liberal institutionalists in IR) alongside 'less formalized norms, rules and procedures that pattern behavior without the presence of written constitutions or material power' (Lipschutz, 1997, p. 96). The latter typically involve 'alliances between coalitions in global civil society and the international governance arrangements associated with UN system' (ibid.), and thus global governance is a multitiered pattern of governance, where actors in (global) civil society are important in generating the norms and rules on which practices are based. Lipschutz emphasises the network character of such patterns of governance; that they are based not on the hierarchy associated with the state but on horizontal relations among a variety of organisations. 'Rather than via global hierarchy or markets, nature will most likely be protected via governance through social relations…in which shared norms, cooperation, trust, and mutual obligation play central roles' (ibid., p. 98). He also emphasises that such patterns of governance are functionally specific, that they emerge in response to particular problems, and that their primary mode of interaction is knowledge-based. The purpose of such networks is learning about what forms of governance 'work'. 'The fundamental units of governance are, in this system, defined by both function and social meanings, anchored to particular places but linked globally through networks of knowledge-based relations' (ibid.).[2]

Lipschutz, however, contextualises his account of emerging patterns of global environmental governance rather differently from Wapner. He places these developments much more clearly in the context of large-scale shifts in the global political economy. He characterises these as 'economic integration accompanied by political fragmentation' (ibid., p. 84). Lipschutz argues that the primary significance of these twin, connected developments is to intensify the difficulties of achieving effective global environmental governance purely through traditional forms of interstate management (ibid., p. 85). (He also argues later, on pp. 91–2, that shared social meanings, another prerequisite of effective global governance, are effectively impossible to reach at interstate levels – such meanings can only be shared at local levels, or among networks of shared interests). The combined effect of these two shifts is to produce a 'neo-medieval' form of world order, where there are multiple levels of authority and governance from the local up to the global (and many of the 'global' levels are not organised spatially as this metaphor suggests), centred on functionally specific networks of organisations.

Lipschutz gives a number of examples of such networks at work in global environmental politics. He discusses, for example, the activities of the Climate Action Network (CAN), the Global Rivers Environmental Education Network (GREEN), and the River Watch Network (ibid., p. 88), and later, campaigns over the Mattole watershed in northern California (p. 93), the Amazon rainforest (pp. 93–4), Love Canal (p. 94) and by the residents of Owens Valley in Eastern California against Los Angeles (p. 95). He suggests that networks and campaigns such as these constitute the primary site of global environmental governance. But in his conclusion and general conceptual argument, he loses an important element of (at least some of) such networks and campaigns, namely an element of struggle and conflict. Global environmental governance is thought of as a process of learning, where actors develop shared meanings and norms to deal with problems, and locally embedded communities share knowledge and norms through global networks. Lost here is a sense of *who* is involved in such processes. Implicitly, it is actors in (global) civil society, but it is unclear whether state and corporate actors are involved. If not, then how do the governance mechanisms affect the practices of those actors? If they are, then surely conceptualising the process as simply one of learning is inadequate. Either way, then, environmental movements and their allies in civil society in these networks are necessarily involved in some form of struggle with state and corporate actors, as is indeed highlighted in Lipschutz's empirical stories.

This point would be emphasised if we also bring in another consideration which is missing from the otherwise very useful analyses of Wapner and Lipschutz. As already alluded to in the discussion of Wapner's book, one absence is a sense of the substantive outcomes to be produced or promoted by such emerging governing mechanisms. There is a sense in both writers' work that such outcomes are radically different from those produced by prevailing political institutions (growth, centralisation, globalisation, and so on). But if that is the case, the likelihood of such governance being effective in helping to produce sustainable outcomes is dependent on the degree to which they are successful in devising strategies which resist the dominant logics of states, capital, big science, bureaucracy, etc. While learning processes among networks are clearly important, and could extend to parts of states and some corporate actors, it is unlikely that this will be sufficient (at least if the structural argument I have offered throughout this book is accepted). Such actors have entrenched reasons not to learn to 'tread lightly on the earth' (or whatever other catchphrase is used), but rather have strong reasons to resist the emerging patterns of governance outlined by Wapner and Lipschutz. Indeed, they perhaps have strong reasons to promote forms of global management which Lipschutz suggests is unlikely to emerge (1997, p. 85), and perhaps are already doing so, as emphasised by some writers on UNCED (Chatterjee and Finger, 1994; Hildyard, 1993; Shiva, 1993; Paterson, 1996b).

Globalisation and global environmental politics

I will develop these two points, concerning the substantive outcomes generated by patterns of governance, and practices of resistance to existing political forms, below. I now turn to a discussion of globalisation. Partly this is because, as Lipschutz rightly points out, such a context is important for understanding emerging patterns of governance, and ways that such governance can be further developed. But it is also because I would conceptualise such changes rather differently from Lipschutz, and this has important consequences for the development of the argument.

For Wapner, contemporary social change at a global level is conceptualised primarily in terms of the emergence of a 'global civil society'. The basis for considering social interaction transnationally exists, and social actors moving across state boundaries are able to effect political change. Without disputing this, it is perhaps also worth emphasising that this is a very particular, pluralist, account of such global social change, especially in the context of the point that this literature downplays the political conflicts involved in such non-state 'global environmental governance'. Specifically, the emergence of global civil society should be thought of as one of the facets of the broader process of globalisation. Lipschutz does emphasise this context, but I want to engage with work on globalisation in order to develop the argument in ways different to his.

Lipschutz's account of contemporary social change, as 'economic integration accompanied by political fragmentation', resonates with a widespread literature on what has come to be known as (economic) globalisation. Debates about globalisation typically focus on the one hand on whether globalisation is happening (or whether the current shifts in the global economy are better characterised as 'internationalisation', 'triadisation', 'regionalisation' or some other phrase), and the implications of such changes for conventional accounts of the possibilities of political action.[3] Specifically, such debates are often concerned to examine whether globalisation has created a situation where there is no possibility for states to pursue any path of political-economic management other than the neoliberal one dictated by global finance and Transnational Corporations. Lipschutz's account (1997, pp. 86–7) of what Phil Cerny has called the 'competition state' (1990) (although Lipschutz does not use this phrase), operating at the level of local political institutions, is clearly consistent with such concerns. Lipschutz offers an argument which sides with those who suggest that globalisation does in fact attenuate state autonomy and proscribe certain forms of action; this is part of his explanation for why conventional interstate collective action on global environmental change is increasingly difficult to achieve.

Lipschutz also shares with many in debates on globalisation an account of what it is. Although he only briefly defines it, he suggests that 'global economic integration is a condition whose origins are to be found in ... the Industrial Revolution, the rise of English liberalism, and the institutionalisation of free trade' (1997, p. 85). This is slightly different from many of the debates on globalisation, particularly in its idealist focus on 'English liberalism', but it shares with most accounts a focus on a set of discrete trends. Most literature on globalisation focuses on the triad of trade, transnational corporations, and finance, as measures of the process. These are the measures by which both proponents of a globalisation hypothesis, and its critics, tend to advance and evaluate their arguments. Globalisation consists, therefore, of a set of empirical measures through which economies have (or have not, depending on your view) become progressively more closely linked, since either 1945, or the early 1970s. Lipschutz's characterisation of integration is consistent with such a view of globalisation.

However, such an account of what constitutes globalisation fails to see capitalist society as an integrated whole. Seen as such, globalisation can be seen less in economic integration, in the sense of greater amounts of GDP being accounted for by trade or Foreign Direct Investment, and more in terms of a broad reorganisation of the power of capitalist elites to global levels.[4] Globalisation therefore can be seen in increased patterns of interconnection between (previously national or regional) capitalist elites across the globe. Such elites are simultaneously public (as in state officials and politicians) and private (TNC executives, bankers), and their increasing interconnections can be seen in the increased intensity of macroeconomic policy coordination, particularly in G7/8 countries, in the variety of forums for elite consensus formation (such as the Trilateral Commission), but also in the variety of private cooperative arrangements, from credit rating agencies through to interfirm alliances in R&D and production. Simultaneously, the process entails a globalisation of consumer culture which embeds an increasingly large number of people's lives in the daily practices of consumption which tie them in, both materially and symbolically, to the fortunes of global capitalism.

As I have suggested elsewhere (1996b), environmental change and politics have provided fertile strategic ground for such globalisation. The processes surrounding UNCED provided many opportunities for TNCs to promote themselves as 'saving the global environment' (Finger and Kilcoyne, 1997). They intensified their organisation of responses to environmentalism through institutions such as the (World) Business Council for Sustainable Development, and while making sure that no mention of TNCs was made in UNCED documents, entrenched themselves as the main legitimate actors in producing responses to global environmental change (ibid., Hildyard, 1993; Chatterjee and Finger, 1994). This 'Greening of the Global Reach' (Shiva, 1993a) has entrenched the power of TNCs both through legitimation and through intensification of their collective organisation as a 'transnational capitalist class'.

Caroline Thomas suggests that such increased interconnections are best conceptualised as 'the process whereby power is located in global social formations and expressed through global networks rather than through territorially-based states' (Thomas, 1997, p. 6). However, I would take issue with the last part of Thomas's formulation. If conceptualised as a social whole, then the opposition of states to (globalising) capitalism is misplaced. I would use the phrase 'in addition to', instead of 'rather than', in Thomas's phrase. Indeed, states can often be seen as agents themselves in promoting globalisation, as Helleiner's work on the globalisation of financial markets (1994; 1995) demonstrates convincingly. If states are part of a broader social whole (capitalist, patriarchal, technocratic), then there is no reason why state managers and elites will oppose processes of globalisation; indeed there are good reasons to believe they would promote it, since (at least as far as globalising capitalist elites are concerned) it promotes the accumulation which is one of the state's structural imperatives.

This account of globalisation helps to illustrate complications to Lipschutz's argument concerning processes of 'global environmental governance' which I outlined above. If globalisation is a central process in both capital accumulation and state reorganisation, and capitalism and the states system are necessarily anti-ecological (as I argued in Chapter 3), then appropriate ecological responses to globalisation are ones of resistance, not of (or perhaps more precisely in addition to) learning. Perhaps more can be said about globalisation here than simply that it is an expression of the logic(s) of capitalism and the state system. Globalisation intensifies existing dynamics of capitalism which tend to disrupt ecological systems. This is primarily through distanciation – that as globalisation increases the physical distance between producers and consumers, it makes it increasingly difficult to be aware of the ecological or social consequences of one's consumption practices.[5]

The appropriate forms of political action to respond to global environmental change in this political-economic context are therefore resistive. This could be seen to have two aspects. Firstly, it can be seen as resistance to globalisation. Since the world's structures of power are fundamental sources of (both) globalisation and global environmental politics, resistance to globalising processes can be part of broader resistance to the structures themselves. Secondly, it can be seen as resistance within globalisation. Certain aspects of globalisation can perhaps be useful to resisting capitalism, the state, and so on, and to furthering a Green social transformation. For example, the widespread use of telecommunications for global networking purposes by critical social movements can be seen as one of globalisation's unintended consequences, one of the 'chinks in its armour' where possibilities of resistance are created. Paul Preston, President of McDonald's UK, alluding to the McLibel case, suggested that 'one downside of globalisation may be that local incidents soon become international crises' (Vidal, 1995a). Central to the possibilities of this have been the creation both of international media, but also (and perhaps more importantly) alternative sources of global information flows, principally the Internet.

Newell (forthcoming), does place discussion of governance in precisely this context. He shows how, given that globalisation constitutes a reorganisation of world politics so that TNCs can further their interests, involving the retreat not of the state so much as of regulationist strategies by states, NGOs increasingly step in to regulate, or 'govern', the practices of TNCs. NGO strategies are varied, from occasional alliances with TNCs to promote 'learning' as implied by Lipschutz's model, through to shareholder activism, consumer boycotts, counter-advertising, and developing codes of conduct. Through such a lens, it becomes clearer that such governance is often more conflictual than is implied by Lipschutz.

Resistance and transformation

My main contention in this section is that the processes of global environmental governance which Lipschutz discusses are perhaps most fruitfully thought of as processes of resistance.[6] Such resistance is to an extent about holding those with power in the global economy, and/or in states to account, making them legitimise their actions, democratising them, transforming their effects. It is in this sense a form of governance as outlined in the global governance literature. But it is also about (re)creating fundamentally different sorts of social space and political economy. It is not only about regulating, governing the practices of TNCs, but also about creating spaces where TNCs cannot dominate. Lipschutz gives some examples which do fit into this category, but a whole host of others could be given. In Chapters 5 and 6 I discussed some in relation to cars and road-building and in relation to the fast-food industry.

A classic example of such resistance in this literature concerns the patenting of seeds and other plant genetic material (e.g. Guha and Martinez-Alier, 1997, pp. 109–127; Shiva and Holla-Bhar, 1993; Kneen, 1995). Numerous other examples could be given, from the Chipko movement, which is universally mentioned, but whose meaning is heavily contested (see, for example, Bandyopadhyay and Shiva, 1987; Weber, 1988; or, for a perspective critical of the conventional view of Chipko, Rangan, 1996), to the environmental justice movement (Bullard, 1990; Szasz, 1994). Taylor's edited volume *Ecological Resistance Movements* (1995) presents perhaps the broadest account of such resistance. Recognising the inevitable diversity of such movements, Taylor nevertheless suggests that there are some features connecting those discussed in his book, from all over the world. He argues (1995a) that they have three distinctive features. Firstly, such movements respond to threats to livelihood and survival; 'popular ecological resistance often originates in a desperate quest for survival as industrial processes threaten habitual modes of existence and as people recognize that their well-being is being threatened by environmental degradation' (Taylor, 1995a, p. 335). Secondly, they involve specific attempts to preserve or create ethical sensibilities towards the non-human world and within human societies conducive to producing sustainability. Taylor terms such shifts in consciousness 'moral and religious' (1995a, p. 336). Such motivational concerns often connect deeply with material interests in survival to produce these movements. Thirdly, along with many other commentators (e.g. Seager, 1993; Mellor, 1992), Taylor notes that in many of the movements discussed in his book the gendered impacts of environmental degradation result in ecological resistance movements very often being women-led.

Gramscian notions of counterhegemony are perhaps one conceptual tool to understanding such resistance. Some, for example Seel (1997), have explained roads protests in the UK in such terms. But the usage of counterhegemony in neo-Gramscian IPE remains rooted in an understanding of politics in terms of reviving social democracy in the face of a neoliberal project which has globalisation as its main rhetorical device (see Paterson, 1999, for more detail). When ecology is dealt with by Gramscians in IPE, it is usually dealt with in a cursory fashion, clearly as an add-on extra which doesn't challenge their main political purposes (e.g. Cox, 1999, p. 9; Gill and Law, 1988, pp. 370–4).

Polanyi's (1957) notion of a counter-movement is perhaps more fruitful. As Bernard (1997) emphasises, Polanyi understood the counter-movement to be a reaction to what he termed the 'disembedding' of the economy, as social forces promoting market freedom successfully disengaged the economy from webs of social obligation. Central to such disembedding is the ongoing creation of what Polanyi called the 'fictitious commodities' of land and labour (fictitious in that they are not produced by the market). The counter-movement arises because people object to the way in which the commodification of land and labour disrupts their capacities to meet subsistence needs, as well as because they recognise that such disembedding is dangerous in the context of the necessary human interdependence with the rest of nature. So for Polanyi, resistance is directly connected to the ecological disruptions produced by a liberalising capitalism, thus conceptualising resistance similarly to the way Taylor characterises empirical examples of ecological resistance (see Bernard, 1997, for an extended account of Polanyi, and also Mittelman, 1998, for a similar application to ecological resistance movements).

At the same time, Polanyi's account of the goals of resistance is also more immediately consistent with a Green concern than is a Gramscian account. For him, the reembedding of the economy meant much more than simply taming the market with rules, as in Ruggie's use of Polanyi in describing the Bretton Woods system as 'embedded liberalism' (1983; Bernard, 1997, p. 86). It involves precisely 'removing the market as the dominant institution in society' (Bernard, 1997, p. 86).

This is much like the arguments of an increasing number of writers on development who call for the 'end of development' (Escobar, 1995, p. 19), which I outlined in Chapter 3. At that point I effectively implied that this was simply an academic trend. But, as Escobar emphasises, the rejection of development is also made by many grassroots movements, especially (but not solely, as we have seen in Chapters 5 and 6) in the South (ibid., pp. 215–16). These authors and movements are 'not interested in development alternatives but in alternatives to development, that is, the rejection of the paradigm altogether' (ibid., p. 215).

The practices of resistance just described, as well as those discussed in Chapters 5 and 6, should therefore be conceptualised in this context.[7] Of course, as Taylor emphasises (1995a) they are highly diverse, and to make this move here is highly problematic. But even where they are not explicitly ecological in intent, these movements (as discussed by Escobar and others) reject fundamentally the dominant social logics which, as I have tried to show in Chapter 3, are inevitably ecologically unsustainable, and work towards social systems which are more conducive to sustainability. Taylor suggests that they are often explicit in understanding 'the problem' in broadly these terms – that environmental degradation and social disruption result from the incursions of global capital and national economic and political elites, that economic growth is not an answer and is impossible in the long run, that commons produce political-economic mechanisms conducive to social stability and environmental sustainability, and that political and economic decentralisation is necessary to achieve these goals (1995a, pp. 337–41). Chapter 3 showed how development is a central discourse through which patriarchal, technocratic, capitalist, statist societies reproduce themselves. By opposing development, such movements oppose the logics of those structures.

Escobar emphasises this point (although perhaps rather differently) through poststructuralist lenses. By focusing on development as discourse, or 'regime of representation', he shows that opposition to development is fundamentally cultural (in the broad sense). It opposes the forms of knowledge and ways of knowing, the techniques and strategies of power, which are epitomised by and practised through development as a discourse, and aims to construct societies not based on such practices of domination. As he argues, while highlighting the importance of ecology in such struggles:

> These struggles – between global capital and biotechnology interests on the one hand, and local communities and organizations on the other – constitute the most advanced stage in which the meanings of development and post-development are being fought over... [they] raise[s] unprecedented questions concerning the cultural politics around the design of social orders, technology, nature, and life itself.
>
> (Escobar, 1995, p. 19)

An account of development as discourse has many similarities with the account of globalisation given above. If globalisation is about the reorganisation of capitalist power to global levels, then such a shift is necessarily discursive. It involves shifts in the forms of power and legitimation engaged in by capitalist elites, new means through which such power is reproduced. Such an account of globalisation is given both by Ian Douglas (1997), and by J.K. Gibson-Graham (1996). For both of these writers, globalisation as discourse operates in order to bring the effects it purports to describe into being. It does so particularly effectively since a central part of its discourse is that such processes are inevitable, irresistible. Gibson-Graham suggests this argument by analogy with the 'rape script': just as discourses of rape tend to render women as powerless victims, which becomes self-fulfilling, discourses of globalisation render societies as powerless to prevent the 'penetration' of global capital. Thus, globalisation should be understood as a 'language of domination, a tightly scripted narrative of differential power' (Gibson-Graham, 1996, p. 120). Douglas is, in a similar vein, primarily concerned with:

> the way in which a series of social imperatives has been established on the back of the rise to hegemony of the concept of globalisation. These imperatives include: 'agility', 'rapidity', and 'mobility'; 'transformation', 'adaptation', and 'invention'; 'competitiveness', 'outlook' and 'foresight'; 'self-reliance', 'self-motivation' and 'self-monitoring'; 'economy', 'efficiency' and 'excellence'; the list continues.
>
> (Douglas, 1997, p. 165)

Through such imperatives, globalisation is 'seen to be inexorable (a logic to which "there is no alternative") and inevitable' (ibid., p. 166).

Douglas's account in particular shows many similarities with the account given of development by Escobar and others. Both involve mobilisations of people in new ways to meet needs of accumulation. Both involve new forms of knowledge and techniques of power, through which subjects are produced as efficient workers. Douglas indeed shows how such mobilisations are contemporary expressions of prevalent forms of power in modernity.[8] Globalisation and development are from this perspective different aspects of dominant discourses through which social structures are reproduced. Thus in the concrete examples discussed above, resistance to development and resistance to globalisation amount to much the same thing, and these practices therefore are direct analogies (in the terms of my argument) to Lipschutz's networks of global environmental governance.

Toward a sustainable world?

But where is such resistance heading? It should be emphasised that resistance is, as I hope is obvious from the previous section, simultaneously resistive and reconstructive. As Esteva puts it (1992, p. 20), such resistance (usefully defined by him as a struggle 'to limit the economic sphere') is seen by its practitioners as 'a creative reconstitution of the basic forms of social interaction, in order to liberate themselves from their economic chains'.[9]

The predominant image within Green literature, emphasised in many of the instances of resistance given above, is of the (re)creation of small-scale, anarchistic, egalitarian communities. This can be seen, for example, especially in The Ecologist's (1993) invocation of the 'commons' as the normative site both of resistance and of a sustainable social system, in Sachs' (1997) notion of the 'home perspective', in Latouche's (1993) concept of the 'informal', in Mellor's (1995) inversion of a conventional phrase concerning political strategy 'getting here from there', and in many other formulations, perhaps most classically in the slogan 'Think Globally, Act Locally'.

The localist imagery and focus is also prevalent in the 'anti-development' literature. Escobar, for example, emphasises such concerns in resistance movements (1995, p. 19, pp. 222–3). 'The nature of alternatives ... can be most usefully gleaned from the specific manifestations of such alternatives in concrete local settings' (ibid., p. 223). He suggests that one of the two defining features of such resistance is 'defense of cultural difference, not as a static, but as a transformed and transformative force' (ibid., p. 226), cultures which he locates not in nation-states but in local communities. For him, the emphasis on the specificity of localities is in part because of a rejection of the dominant development paradigm as inevitably universalistic, and of such universalisms as inevitably dominating. 'To think about alternatives in the manner of sustainable development, for instance, is to remain within the same model that produced development and kept it in place' (ibid., p. 222).

In many ways, these two literatures – Green, and anti-development – emphasising the (re)construction of local community-based social and political forms, have much in common, albeit coming to the question from differing backgrounds. Esteva and Prakash (1997) are a good example of these two backgrounds coming together in the same place. For reasons both to do with the failure of development, and to do with the imperatives of ecology, they emphasise localism. For them, this involves necessarily also emphasising not only a localism of action, but also of thought. They take apart the slogan 'Think Globally, Act Locally'. This suggests that a form of global consciousness is necessary, but it 'rejects the *illusion* of engaging in global action' (ibid., p. 278), both because it is impractical, but also because it involves an 'arrogance', a 'far-fetched and dangerous fantasy' (ibid.). Global action will necessarily be oppressive, managerialist, domineering, the slogan implies. Esteva and Prakash associate such a view with what they term 'alternative global thinkers' (they name James Robertson and The

Other Economic Summit, David Korten, and Greenpeace, ibid., pp. 288–9, note 5). But they also suggest that global thinking is equally dangerous, and that thought should also be local. They suggest firstly that global thinking is impossible:

> To fit the Earth conveniently into the modern mind, the latter has shrunk it to a little blue bauble, a mere Christmas-tree ornament; and invited modern men and women to forget how immense, grand, unknown and mysterious it is'.
>
> (Ibid., p. 278)

Global thinking is thus designated 'God-like'. In practice, they suggest, we 'can only think wisely about what we know well' (p. 279) and thus advocate the 'Wisdom of thinking little', which they associate with Gandhi, Illich, Kohr, Schumacher and in particular Wendell Berry. They cite Berry's arguments concerning food. Berry suggests that if we start with this, there is no need to 'think global' to resist the forces which produce food in an unsustainable and unjust manner. The focus should be to build local alternatives to global food industries. Such local alternatives would have a better capacity to be ecologically sound and based on just treatment of workers, because the conditions of production are visible to those consuming the food (ibid., pp. 279–81).

Kuehls (1996), Dalby (1998b) and Stewart (1997) all suggest that this localism in Green politics is problematic. One older objection to such images is that such small-scale Green communities would rather likely be parochial, inward-looking, and perhaps even xenophobic, and at least that there are no guarantees that they would not simply export their ecologically problematic features on to other communities. Kuehls', Dalby's and Stewart's objections are connected, but put rather differently.

Dalby suggests, while agreeing with Green critiques of 'global environmental management', that:

> The political dilemma and the irony here is that the alternative to global management efforts – that of political decentralization and local control, which is often posited as the political alternative by green theory – remains largely in thrall to the same limited political imaginary of the domestic analogy, and avoids dealing with the hard questions of coordination by wishing them away ...
>
> (1998b, p. 13)

The second part of this critique, concerning questions of coordination, is relatively conventional. Critiques of localism within Green theory, for example by Eckersley (1992) or Goodin (1992) are very similar. I have discussed these already in Chapter 3 (and at more length in Paterson, 1996c).

The earlier part of Dalby's critique, however, is that Greens remain committed to a sovereign model of politics, the 'domestic analogy'. This, although put in a different theoretical context, is the same form of critique as made by Wapner, discussed earlier in this chapter. As I stated there, it is a mistake to suggest that Green arguments for decentralisation remain within a sovereign model of politics. Shifts in the scale of political organisation are simultaneously shifts in forms of political organisation.

Kuehls does recognise that Bookchin, whom he discusses as an exemplar of this Green localist position, is anti-state (it would be difficult to avoid this conclusion). But somehow Kuehls implies that Bookchin remains committed to the state as a model of political organisation, as the site of politics:

> Ophuls's and Bookchin's theories are easily placed onto O'Riordan's and Dobson's matrix of ecopolitical thought due to their similar orientation to state and sovereignty. Although one [Ophuls] endorses the state as an appropriate place for ecopolitics and the other holds the state to be the absolute negation of an appropriately ecological politics, both reify the state as the locus of political activity – for good or bad.
>
> (Kuehls, 1996, p. 106)

This conclusion seems perverse to me. Bookchin is held to negate the state but to be committed to it as the locus of political activity. His anti-state position must therefore be an anti-political position, suggesting that in his 'utopia' (for want of a better term) there would be no politics. This seems a bizarre reading of Bookchin, who would be entirely happy with the notion that politics occurs outside the realms of the state, as Kuehls (and others discussed in this chapter) assert. A model of politics, like Bookchin's 'municipal confederalism', which rejects the state, necessarily rejects sovereignty, and is therefore open to possibilities for global coordination which Dalby and many others imply is necessary.

The other part of Kuehls' critique of green localism is that it fails to appreciate 'the difference of political space that exceeds the sovereign territorial state model of political space' (Kuehls, 1996, p. 106). By this,

Kuehls refers on the one hand to Haraway's notion of the cyborg, which engenders a society which has moved 'from comfortable old hierarchical domination to scary new networks…of domination' (Haraway, 1991a, as cited in Kuehls, 1996, p. 108). On the other hand, although clearly connected to the former, he refers to the obsolescence of the 'sovereign territorial state model' in terms of economic globalisation (although he doesn't use the phrase). Citing Harvey (1990), he discusses the capacity of global finance to switch flows of capital around the world at speed as evidence here. 'The old model of the body politics is inadequate in the face of the networks of global capital' (Kuehls, 1996, p. 108). Drawing on Deleuze and Guattari, Kuehls refers to such a global form of political space in terms of 'nomadic trajectories and worldwide machines' (ibid., passim).

But such a characteristic surely takes the current trajectories of global politics and implies that they are irresistible, irreversible. The point of writers like Esteva (upon whom Kuehls draws extensively in his account of governmentality) is precisely to suggest that to the extent that globalisation is occurring, it should be challenged and countered. The 'worldwide machines' should be resisted and dismantled. The descriptions of contemporary global politics by Kuehls, Dalby, or in a different language but with similar effect, by Wapner and Lipschutz, are premised on the continued existence of a globalising capitalism, a continued existence which Greens want to prevent. Discussion of political forms need not therefore be premised on such global economic connections.

Of course, global ecological flows would still exist in a world of small-scale, self-reliant ecological communities. For most advocates of such a world, so would global social and cultural flows. But this does not invalidate a model of politics which *starts with* the local and looks outwards, rather than one which starts with the fact of transnational economic flows, as do Kuehls, Dalby, Wapner and Lipschutz, and suggests that such flows will always exist as a constant within which political actors will have to operate.

Ultimately, these critiques conflate two notions which should be kept separate. Sovereignty is conflated with any concern for a form of politics which is connected to place, that is, to communities one of whose constitutive features is their situatedness in particular places. Stewart makes this move most clearly (1997, p. 13), in discussing The Ecologist (1993):

> Once again the assumption is that political containers are appropriate as long as they are locally controlled, and democratic.

> Sovereignty is a problem if it means illegitimate use of power…but sovereignty for the people in a commons is ethical. Local communities ought to have the power to decide their fate, the problems are a matter of external threats and violations of these rights. Once again the principle of state sovereignty lurks in these ideas of commons and locality. Political life once again is reduced to a matter of living in unique local boxes.
>
> (Stewart, 1997, p. 13)

The conflation here suggests that any argument for living in spatially discrete places and communities must of necessity also involve an argument that these 'political containers' or 'local boxes' must be sovereign – they must regard themselves as having exclusive rights to determine what goes on in that place. In my view, this is an implausible interpretation of most of the writers whom Kuehls, Dalby and Stewart discuss. Localism need not be committed to a sovereign model of politics even while it is committed to building communities which have much more intense connections to particular places than do societies in modernity.

As Helleiner (1996) points out, such arguments perhaps miss the meaning of the most popular of Green slogans, 'Act Locally, Think Globally'. Helleiner emphasises that the initial usage of that phrase by Dubos[10] was precisely to guard against parochial localism, by emphasising the way that local action is embedded in global modes of consciousness and normative beliefs. In addition, as Esteva and Prakash point out, it is precisely global proposals which are parochial:

> Global proposals are necessarily parochial: they inevitably express the specific vision and interests of a small group of people, even when they are formulated in the interests of humanity.
>
> (1997, p. 285; see also Shiva, 1993a)

As Esteva and Prakash emphasise, a rejection of global thinking does not mean that resistance, and the communities that resistance constructs, are not thought of as connected. One objection to an emphasis on local action is that such initiatives 'seem too small to counteract the "global forces"' (1997, p. 281). They suggest that local resistance does indeed need outside solidarity and allies. But they reject the idea that this necessarily involves 'thinking globally':

> In fact what is needed is exactly the opposite: people thinking and acting locally, while forging solidarity with other local forces that

share this *opposition* to the 'global thinking' and 'global forces' threatening local spaces.

(Esteva and Prakash, 1997, p. 282)

They suggest that the magazine *The Ecologist* is a prime example of such a political practice. While much of its work demonstrates and supports ecological resistance around the world, it refuses a 'God-like' global viewpoint. 'In no way do these forms of transnational sharing transmogrify local people into globalists' (Esteva and Prakash, 1997, p. 288).

Where does this leave us regarding where I started this chapter, with notions of 'global environmental governance'? Certainly, Lipschutz and Wapner share with Sachs, Esteva and others a rejection of a 'global management' image of such governance. But they do not see such governance as resistive of dominant social forces. The networks, movements, which make up global civil society, and which practise governance in that society, are separated in Lipschutz and Wapner's arguments from the practices they govern. Those practices (I assume they are not explicit) are those of TNCs, governments and perhaps consumers. As Samhat puts it, such an approach focuses on 'civil society networks that raise consciousness, mobilise social forces, monitor behaviour, and expose transgressions' (1997, p. 378). All of these roles presume that it is someone other than the members of the networks who is being governed. In the models of governance envisaged under the localist visions of *The Ecologist* and others, there is no such separation. Political economy becomes again precisely that; it rejects the separation of polities from economies that is fundamental to capitalist modernity. While Lipschutz and Wapner's arguments represent perhaps the best which International Relations has yet come up with regarding thinking seriously about political responses to the ecological crisis, a more radical vision is nevertheless needed. The notion of networks in global civil society is a useful way to think about how and where action to promote sustainable societies might occur. But for the practices of such networks to actually promote sustainability, their radicalisation towards local and solidaristic resistance practices and construction of alternative forms of political and social life, are necessary.

The fundamental point here in terms of this argument of this book as a whole is that the question of governance, whether localist or 'rhizomatic', should be informed by a conception of the origins of global environmental change. Like liberal institutionalists or realists, Kuehls and others only tangentially address this question, perhaps eschewing such notions of causality as overly modernist. Yet a set of assumptions about such causes is a necessary part of evaluating these debates about forms of politics appropriate to producing a sustainable society. I have tried to show that global environmental change is driven by twin dynamics of domination and accumulation, systemically produced by global political structures. Such dynamics of modernity inevitably 'tear space away from place'. A rhizomatic ecopolitics is founded ultimately on accepting such a dynamic, rather than resisting and transforming it. While such networks may help to move societies away from the large-scale processes and structures driving environmental destruction, their effect can never be more than tactical. By contrast, a localist Green perspective is founded on an understanding of global environmental politics that broadly locates the origins of global environmental change in the dynamics I have identified, and this is for me the most appropriate political response to global environmental change.

Feminists make a separate, and perhaps more salient point, concerning the development of local communities (for example, Mary Mellor, 1992). Noting that the arguments made by many Greens for localism are based on some sort of naturalism ('natural' ecological boundaries for bioregionalists, for instance) she argues forcefully that such a politics needs to avoid such naturalisms, and to base arguments on explicitly political principles. Otherwise, the possibilities for reproducing patriarchy in small-scale communities, a frequent feature of such 'natural' communities, cannot be prevented (1992, ch. 4, also pp. 234–8).

Mellor couches some of this critique in terms of balancing local organisation with global forms of politics. In this context she seems uncertain as to how to resolve this local–global dilemma, citing Green decentralist arguments in response to traditional centralising socialist tendencies, and the need for global forms of 'administration' when discussing Green localist arguments (compare, for example, her treatment of it at pp. 113–15, 234, and 238). Her most representative position is that 'we can unite globally only if we break up the nation-state into smaller self-managing communities. A true internationalism can be built only from the bottom up' (1992, p. 234).

But her more important point is that 'we must be careful not to fall victim to the "structural fallacy" – that is, assuming that if the structure is right (be it collectivised control of the means of production or decentralised small-scale communities) the rest will fall into place' (Mellor, 1992, p. 238). I read this as meaning two things. Firstly, it indicates that we need to specify more closely the political basis on which communities should be organised. Secondly it indicates that ultimately there is no substitute for politics – even a 'Green utopia' will involve continual struggle over rights, justice, distribution questions, what forms of production are sustainable, and so on.

Mellor concentrates on the first implication. She suggests persuasively that responding to global environmental change requires a politics which starts from the provision of basic needs. This is also for her what connects Green politics, socialism and feminism (1992, p. 239). Taking basic needs as primary requires that we start from the reality of women's lives in patriarchal societies. This is because it is women's lives which produce most of the basic needs for all people. Men's lives are primarily organised around accumulation-based practices. Basic needs are defined by her, drawing on Max-Neef (in Ekins, 1986, p. 49) as 'subsistence, protection, affection, understanding, participation, leisure, creation, identity (or meaning) and freedom' (Mellor, 1992, p. 239). Green communities would therefore need to be based on the primacy of the values of nurturing, care, and emotional labour, through which such needs are met. At the same time, such communities need explicitly to reject the essentialisms by which such values are assumed to be biologically female, an assumption which reproduced women's subordination. Recovering what Mellor calls the 'WE' world (Women's Experience) will entail continued struggles for women's 'control over their bodies both sexually and in terms of reproduction' (p. 276), and for economic independence from men (pp. 276–7). Men will have to abandon 'the benefits and constraints of both patriarchy and masculinity', to live in the WE world (p. 277).

There are many questions concerning both the contours of such a society and how we might get to it, which are beyond my current capacities to answer. But such a set of basic principles concerning the scale of human societies and the principles on which those societies might be based, are consistent with the argument I have tried to make throughout the book. Since global environmental change is produced systematically by a globalising patriarchal capitalism, we need to work toward societies which *as systems* do not necessarily generate such degradation. They may produce some forms of environmental degradation but this would not be systemic by nature, nor global in scope. And to avoid such consequences, societies must avoid making domination and/or accumulation their basic principles, but rather be based on egalitarian, non-hierarchic politics, and on economics oriented towards meeting basic needs for all.

Notes

1 Introduction: Understanding Global Environmental Politics

1 For example, see Conca, Alberty and Dabelko (1995), Vogler (1995), Haas, Keohane and Levy (1993), Bartlett, Kurian and Malik (1995), Elliott (1998), Chatterjee and Finger (1994), Miller (1995), The Ecologist (1993). Choucri (1993) achieves the same effect, as the foreword is written by Maurice Strong, UNCED's Secretary-General.

2 The formulation here is not intended to exclude the questions thrown up by poststructuralist perspectives. Doran (1995) for example, argues that a critical global environmental politics should be based on poststructuralism. For him, the most pertinent questions are those of power/knowledge. I would argue that my three questions can all be investigated through such a lens. But I would want to make claims concerning the origins of global environmental change central to a critical global environmental politics. Doran makes answers to such questions a matter of rhetoric, not amenable to detailed explanation (such a notion being overly modernist, I suspect). Thus while his normative position is very close to my own, he has little time for, and few conceptual tools for, an analysis of the origins of global environmental change.

3 For example Conca (1993), Dalby (1992), Doran (1995), Elliott (1998), Kuehls (1996), Litfin (1994), Paterson (1995, 1996a), Runyan (1992), Saurin (1994, 1996), Boardman (1997), Helleiner (1996), Hovden (1998), Laferriere and Stoett (1999).

4 I develop this point further in Paterson (2000a).

5 Saurin (1996, p. 85). I take issue with Saurin's argument in one respect. He suggests a classic Marxian focus on production relations. While I agree that these are crucial, dynamics of consumption practices are also important, and while it is clear that they are connected to production, I would argue they are not simply epiphenomenal. For the case here, it is more that the direct consumer is the 'agent' of environmental change than the direct producer (Saurin, 1996, p. 87). I discuss questions of consumption in more detail in Chapter 6.

2 Realism, Liberalism and the Origins of Global Environmental Change

1 Also on effectiveness, see Keohane, Haas and Levy (1993), Bernauer (1995), or Victor *et al.* (1998).

2 For others who use this threefold typology of power, interests, and ideas, see, for example, Rowlands (1995), albeit in a slightly modified form, Vogler (1992), or Hansenclever, Mayer and Rittberger (1996).

3 CERES stands for the Coalition for Environmentally Responsible Economies. They were formerly known as the Valdez Principles. Other examples could be given, such as the Forestry Stewardship Council, established by WWF to certify timber that is sustainably produced (Humphreys, 1996, pp. 149–51). In a more conflictual manner, organisations like Multinationals Monitor and even magazines such as *Ethical Consumer*, or campaigns like those against McDonald's or Nestlé, fulfil such governance functions. See Newell (forthcoming 1999). This question is also discussed further in Chapter 7.

4 While in many ways the political dynamics of the ways environmentalists have tried to advance an environmental security agenda which disrupts the traditional focus of security in nationalist/statist and militarist terms is highly interesting, and reveals much of the dynamics of using existing political institutions to achieve Green goals, it is not the focus of this section. For examples, see Deudney (1990); Mische (1989); Finger (1991); Dalby (1992, 1998b). For a useful overview of the debates, see Kakonen (1994), and for selections of key texts, see Conca, Alberty and Dabelko (1995, pp. 239–77) or O'Tuathail, Dalby and Routledge (1998, pp. 179–243).

5 Waltz writes that 'The death rate among states is remarkably low. Few states die; many firms do. Who is likely to be around 100 years from now – the United States, the Soviet Union, France, Egypt, Thailand, and Uganda? Or Ford, IBM, Shell, Unilever and Massey-Ferguson? I would bet on the States, perhaps even on Uganda.' (Waltz, 1979, p. 95)

6 In this schema, *I* refers to environmental impact, *P* to Population, *A* to Affluence (sometimes *C* for Consumption is used), and *T* to Technology. Prominent in debates over this in the 1970s was the one between Barry Commoner and Paul Ehrlich over the relative importance of technology or population. The typology provided by Choucri (1993a) cited earlier is clearly a version of this threefold explanation of environmental change.

3 The 'normal and mundane practices of modernity': Global Power Structures and the Environment

1 The phrase in the title here is taken from Saurin (1994).

2 Several parts of this chapter were previously included in my chapter 'Green Politics' in Scott Burchill's *Theories of International Relations* (1996; Paterson, 1996c). These include much of the section on Green political theory, the section 'Against development', and the concluding section.

3 This discussion is perhaps rather arbitrarily centred on these three books. For other discussions of Green political thought, see O'Riordan (1981), Hayward (1994), Atkinson (1991) or Martell (1994).

4 This section will follow Eckersley's *Environmentalism and Political Theory* (1992), largely for reasons of simplicity, but also because her book still represents the most developed application of ecocentric ideas to politics. The references here will simply refer to page numbers from that book. For other ecocentric works, see, for example, Birch and Cobb (1981) or Fox (1990). Hayward (1994), for one, is sceptical that a full ecocentric position is necessary for a radical ecological politics, and argues for an ecological humanism.

5   The other positions which Eckersley identifies are resource conservation, human welfare ecology, preservationism and animal liberation (1992, ch. 2).

6   The classic early critique was provided by researchers at the University of Sussex (Cole *et al.*, 1973). For many of the raw materials they predicted would run out by 2000 there are in fact now greater reserves than there were in 1972 (reserves being related to price – the higher the price, the greater amounts are recoverable).

7   To see this at work in Green writings, see for example Bunyard and Morgan-Grenville (1987); Porritt (1986); Spretnak and Capra (1985); Trainer (1985); Henderson (1988); Goldsmith (1992). There is an interesting revival of notions of limits in the 1990s, after the domination of notions of 'ecological modernization' and 'sustainable development' in the 1980s and early 1990s. This can be seen in different ways in, for example, Douthwaite (1992), The Ecologist (1993), Booth (1998) and the 'ecological footprint' analysis of Wackernagel and Rees (1996).

8   This concept was originally used in the *World Conservation Strategy* developed by the International Union for the Conservation of Nature (IUCN, 1980), and popularised by the Brundtland Commission, or World Commission on Environment and Development (WCED, 1987).

9   Peter Doran (1995) has also made a similar argument from a poststructuralist point of view within IR. For Doran, however, while there is this general critical approach, there is perhaps an overriding interest in the knowledge structures of global politics. Yet another starting-point broadly consistent with the approach here would be the growing interest in IPE in the work of Karl Polanyi. See, for example, Bernard (1997).

10   I do not of course intend this to imply a simple endorsement of Bull's view. It is intended only to illustrate that the states system is often regarded to involve a number of social mechanisms regulating how states interact with each other. The formulation of sovereignty as 'produced formal independence' is intended to be consistent with an argument that sovereignty should not be seen as a static 'thing', but rather a 'bundle of rights' pertaining, in the modern era to states, but before that to God (via the Pope), which have evolved over time (see e.g. Litfin, 1997; Weber, 1995; Conca, 1994; Bartelson, 1995). In this context sovereignty should not be seen as something defining the (modern) states system, and its evolution is not necessarily evidence of changes in the dynamics of that system. It is common to conflate sovereignty and the states system, as does for example Conca (1994, e.g. at p. 703). Rather, sovereignty is an expression of how state rulers have legitimised their rule, and its content has changed as the means of legitimation have changed. The states system has therefore been able to retain its fundamental features, as outlined here, irrespective of changes in the context of sovereignty.

11   This is not of course inconsistent with the fact that many companies do consider the ecological consequences of their actions. A more upbeat reading of this relationship would emphasise that many firms have undergone radical changes in their production methods to limit emissions, reduce resource consumption, and so on. Such an interpretation often depends on liberal assumptions of 'consumer sovereignty' – that firms are simply responding to consumer demand for more ecologically benign products and

production processes. In this interpretation, there is (at least implicitly) an unlimited potential for such change to occur. However, if assumptions about consumer sovereignty are dropped (i.e. there is much more room for manipulation of demand by firms), and, more importantly, if limits to the potential for improvement in rates of resource use by firms exist, then limits to the 'greening of industry' must also exist. At any rate, what is not in dispute is that any such greening can only occur within a primary concern for the 'bottom line' – what is at dispute is how much change is possible within this constraint. For readings on the 'greening of industry', see, for example, Welford and Starkey (1996), Schmidheiny (1992), Athanasiou (1996, ch. 5), Karliner (1997).

12   This perspective would also emphasise that the external costs assumed by neoclassical environmental economics, to be 'naturally' external, have in practice been historically produced as external; firms have historically acted to make sure that those costs were not included in their production costs.

13   This is to an extent simply an ecological version of worries many critical theorists had concerning scientific-technological domination and its consequences for democracy and human emancipation (Marcuse, 1964; Ellul, 1964; Habermas, 1968).

14   In the case of the last example, the classic case is that of the siting of hazardous waste sites in the US, which has spawned the large environmental justice movement (Bullard, 1990; Szasz, 1994).

15   In fact, Carter places theoretical primacy on the state as the stabiliser of all these other power relations. This reflects his anarchism, and a concern to counter the economism of many Marxists. He develops an anarchist theory of history elsewhere (1989), focusing on the state as a generator of social change, but perhaps risks simply inverting Marxist categories, producing a politicist reductionism to replace the economic reductionism of many Marxists. If that is the case, then the argument here is certainly that none of these structures can be given ontological primacy. Perhaps interestingly, since the argument I develop has much in common with those developed by writers who are often called 'historical sociologists', Carter's argument is in many ways strikingly similar to Tilly's in *Capital, Coercion and European States* (1990).

16   For Green writers making arguments similar to this see, for example, Bookchin (1982); Spretnak and Capra (1985); Porritt (1986). For feminist writers, see Shiva (1988); Seager (1993); Plumwood (1993); or Mellor (1992). And for Marxists or socialists, see Trainer (1985); Ryle (1988). There is an artificiality in separating these writers into their separate boxes; many regard themselves as falling into at least two of these categories.

17   I present here what seems to me the strongest version of the argument. Some Greens do believe it is more fruitful to try to reconstruct the term development, rather than to reject it. However, they would be equally critical of the forms of development criticised by writers mentioned in this section. A debate about whether to reject or reconstruct notions such as development can easily collapse into a simple terminological dispute which is not particularly important. The important point here is that if development is understood as necessarily involving quantitative growth of the system, greater complexity of technological systems, and increasing economic interconnections across the globe, then Greens are clearly opposed to it.

18   For a selection of writings advancing arguments like the one discussed here, see Sachs (1993b, 1992b), Escobar (1995), Esteva and Prakash (1997), Esteva (1987), The Ecologist (1993), Shiva (1988), Trainer (1989), Apffel-Marglin and Marglin (1990). For other discussions, see Hayward (1994, pp. 14–16) or Kuehls (1996, pp. 81–5).

19   For a critique of Brundtland along these lines, see Visvanathan (1991) or Kuehls (1996, pp. 81–5).

20   This is not necessarily inconsistent with an argument that economic growth (at least under capitalism) necessarily increases inequality. Inequality increases while the poor are bought off through quantitative increases in their own consumption, produced by growth in the system as a whole.

21   Like many other Greens, she suggests that social justice, in the form of at least a considerably more egalitarian world than that which currently exists, is an integral part of ecocentric ethics.

22   Examples of this would be Ophuls (1977) and Hardin (1974).

23   There is an important, although never clearly defined, difference between self-sufficient and self-reliant. The former implies that there should be no trade or other exchange of material resources between communities, while self-reliant is a less strong injunction: merely that communities should be primarily only dependent on their own resources, using exchange with other communities where they cannot produce particular items themselves.

24   Bioregionalists argue that ecological societies should be organised with natural environmental features such as watersheds forming the boundaries between communities.

25   The notion of subsidiarity is often used in Green discourse. It is not, however, used in the way that many governments use it – to protect their rights against those of supranational organisations (the classic case being the UK government in relation to the EU). In the Green version, it has radical implications for decentralisation of power to the local level, with power only transferred to higher levels if deemed necessary – local levels deciding what is necessary.

26   See also Wall (1994).

27   I discuss some objections to the 'ecoanarchist', and 'decentralist' position in Paterson (1996c, 1999a). In particular, Goodin (1992) suggests that some forms of global, authoritative institutions are necessary under any Green scenario. He suggests this on game-theoretic grounds. His primary rationale is that any possible future which we can envisage will involve communities across the globe in ecological interdependencies, and therefore with the need to cooperate with each other. Goodin, however, makes a misleading move when he concludes that this will necessarily require authority to be located at global levels. His game-theoretic models could lead equally to the conclusion that only mechanisms to facilitate communication would be necessary, which is something rather different from authoritative institutions.

28   See also The Ecologist (1993) for extended discussions of similar arguments.

29   It will be clear from what follows that this is a usage of commons which renders the term 'global commons', in widespread use in mainstream environmental discussions, to refer to problems such as global warming or ozone depletion, nonsensical.

30 This has a lot in common with the game-theoretic argument mentioned above which often emphasises how in small-scale systems it is easier to generate cooperation than in large-scale systems. Ostrom (1990) develops this argument explicitly with respect to commons regimes.

4 Space, Domination, Development: Sea Defences and the Structuring of Environmental Decision-Making

1 For a more descriptive history of the construction of sea defences in the Netherlands, drawn on by Schama, see Lambert (1971).
2 According to McPhee, this is for largely historical reasons, when the river had to be protected during the war with the English in 1812. The Army simply kept those responsibilities afterwards. Water is not the only aspect of nature which is treated as an enemy. For just two richly detailed examples, see Rodman's (1993) account of the understanding of 'invasion' by exotic plants in a number of cases in the US, or Lowenhaupt Tsing's (1995) account of the 'killer bees' moral panic in the US. In the latter, direct links are made between the 'invasion' of the US by a new species of bee, and racialised understandings of US identity. For Brazilian scientists, where the bee 'originated', the bee was understood and celebrated as a multicultural hybrid of African, European, and indigenous American strains of bee. US scientists, by contrast, understood this hybridity in terms of danger, reminiscent of US notions of the 'melting pot', and tried to fix the identity of the bees as African.
3 For more on managed retreat, see Pethick (1993). On some of the political conflicts involved in putting it into practice, see *The Guardian*, 29 November 1994, or 4 August 1995; Hutchings (1994), or many of the submissions to the House of Commons Agriculture Committee (1998a; 1998b).
4 Elvin and Ninghu's (1995) discussion of historical changes in a river course in China provides a healthy (if inadequate) corrective to the Eurocentrism here. Although much of the evidence regarding early construction of sea defences is impressionistic, their picture of the construction of such works is impressively similar to the pattern in Europe. Sea walls appear to have been built earlier in China, but groynes were first used (as far as one can tell from their narrative) in the eighteenth century, much as in Europe.
5 This is also reflected in most contemporary policy discussions regarding sea defences. See, for example, Turner, Bateman and Brooke (1992), who use cost–benefit analysis to discuss a case in Aldeburgh, and focus primarily on the number of properties to be affected. Cunningham (1992) provides the only discussion of sea defences by somebody working within political studies (that I could find), and confirms the dominance of cost–benefit analysis and financial considerations, in her case in relation to a technocratic model of decision-making where engineers are held to dominate (she cites Parker and Penning-Rowsell, 1980 as using such an assumption). Also see Viles and Spencer (1995, pp. 289–311), Parker and Penning-Rowsell (1983). In an entirely different context, Elvin and Ninghu (1995, p. 8) suggest that sea defences in China also had more to do with protecting economic functions than protecting lives, although in their case the economic functions to be

protected were agricultural (as in much of rural England or the Netherlands, for example).

6 Groynes are the wooden structures built perpendicular to the coastline to slow down the movement of shingle along the shoreline and therefore keep it on particular beaches. The shingle on the beach is what protects the coast from being inundated by the sea in high tides or storms. Without shingle (and therefore groynes) sea walls are quickly eroded by the sea.

7 Sea defences are covered nationally in the UK under the Coast Protection Act (1949), and the Land Drainage Act (1976). Under the latter, the Ministry of Agriculture, Fisheries and Food pays a substantial portion of the costs of replacing sea defences, but has to approve the scheme according to various criteria. For a history and comparison of the two Acts and problems of compatibility between them (although this discussion does not cover the changes in organisation in the early 1990s), see Trafford and Braybrooks (1983).

8 SGS is a large transnational company concerned with certification and auditing the management systems of companies. SGS Forestry is its forestry division.

9 See the minutes of relevant meetings of the Council's two committees involved, the Strategic and Economic Development sub-committee of the Policy and Resources Committee, and the Environment Committee, contained in Eastbourne Borough Council (1992; 1993; 1994a; 1994b; 1995a; 1995b; 1996).

10 Counsell also attributed the structural narrative underlying the *Channel 4 News* story to political factors. The story (April 1995) was shortly before that year's local elections, and prominent Liberal Democrat MP Matthew Taylor was due to visit Eastbourne (where the Council was Liberal Democrat dominated). Counsell suggested that this inhibited Channel 4 from being more critical of the Council. Interview, 29 November 1996.

11 This interpretation is based on an interview with Peter Padgett, Senior Engineer at Eastbourne Borough Council, 27 November 1996.

12 This publication has no date on it, but must have been produced in early 1994.

13 This is not to suggest that such reasons were contrived to mask the importance of tourism and aesthetics. The Council was open about tourism being important (interview with Peter Padgett, 27 November 1996).

14 Ray Russell is an architect local to Eastbourne with a reputation as a 'Green' architect, who was involved in the campaign against the use of greenheart in the project.

15 That Eastbourne does have such an image can be seen in the Lonely Planet Guide to Britain, whose entry on Eastbourne describes it as 'uninspiring and depressing.... Most (visitors) seem to be elderly pensioners on package holidays. Trying to solve the riddle of why they come can lend a brief, gruesome, fascination' (quoted in Coles, 1995, p. 24).

16 The importance of tourism can also be seen in the way that the Council and the engineers planned the construction of the new groynes. Many of the discussions in Council minutes discuss this question and are keen to minimise disruption to tourism. For example, 'works within "tourist sensitive areas" would be mainly confined to the autumn/winter/spring period' (Eastbourne Borough Council, 1993, p. 744).

17 Ray Russell, letter to Bert Jones (Eastbourne Borough Council Chief Engineer), 3 March 1994, and press release, 28 February 1995.

18 Guyana is, perhaps ironically for this case study, a country which has by and large escaped the worst of deforestation (Myers, 1994, p. 31). Its level of forest cover in 1990 was still 85–90 per cent (Palo, 1994, p. 51), along with Surinam and French Guyana the highest levels in Latin America. On the beginnings of increased deforestation there, see Colchester (1994).

19 Peter Padgett (interview) also stated that when DTL was taken over and the Council was trying to get guarantees from them that they would try to get FSC certification for their greenheart operation, the Guyanan government became involved, resisting the FSC on the grounds of 'eco-colonialism'.

20 For general reviews of some of these factors see, for example, Thomas (1992, pp. 245–54), or Mahur and Schneider (1994). The timber trade is not in general regarded as a major direct cause of deforestation, but logging operations can have important secondary effects leading to deforestation (Barbier *et al.*, 1994).

21 This is, however, exactly the same as produced for the Selsey Bill to Beachy Head Shoreline Management Plan, the next stretch of coastline to the west, so there appears to have been no significant shift in meaning as the definition is translated from central to local government.

22 Other participants in the process in Eastbourne reflect such a notion of sustainability. In Posford Duvivier's brochures it has the same meaning, and excludes consideration of the sustainability of the production of the materials used in projects (Posford Duvivier, n.d. (a), n.d. (b), received in 1997). In their 'Environment' brochure, there is a photograph entitled 'Assessment of forest management practices, Guyana' (n.d. (b), p. 2), but no mention in the text that this is part of their understanding of sustainability. Similarly, the Environment Agency's section on 'Environmental issues' in its Principles of Floodplain Protection, exclude such considerations (Environment Agency, n.d.), as do Aitken and Howard in their materials on Greenheart (n.d., received 1997) and the House of Commons Agriculture Committee (Agriculture Committee, 1998a).

23 See for example editions of the *Eastbourne Herald* on 26 November 1994, 11 March 1995 or 15 November 1995, for examples of this representation.

24 Posford Duvivier and the Council did make many commitments and claims concerning DTL's commitments about the social effects of their forestry practices, that they would take such considerations into account. In particular, claims were made that at Mabura Hill, the site where DTL had a concession to log greenheart, no indigenous people were living. However, at the point at which the debate was scientised (as I suggest was the focal point of the debate), such considerations disappear from view.

25 An interview with councillors involved (Janet Grist and Mrs M. Pooley, Chair and Deputy Chair of the Environment Committee of the Council) showed, in a contradictory way, both that they deferred to 'their' experts, the Council's officers and consulting engineers, in matters of scientific expertise, rejecting FoE evidence in consequence, but could also speak quite articulately on the matter of the evidence provided by the officers. Interview, 19 February 1997.

5   Car Trouble

1   The local Conservative MP, Adrian Rogers, advocated starving the protestors, depriving them of air and water, and using CS gas on them, according to one of the protestors' father, Addrian Hutson ('Letters', *The Guardian*, 8 May 1997, p. 18).

2   For more details on the construction of Swampy and the other protestors on the A30 protest in the UK press, see Paterson (2000b).

3   When I gave a paper on this topic at Carleton University in April 1999, it was written up (favourably) in the 'Wheels' section of the *Ottawa Citizen*, as part of a regular column entitled 'Clearing the Air' (23 April 1999, C4).

4   Other car manufacturers have also espoused the joys of unnecessary car-driving in their ads. VW, for example, in early 1999, has billboard ads for the Bora, which says 'Distant Relatives Beware', and then has a slogan under the VW logo and the name of the car, 'Any Excuse'.

5   The force of this advert may have been helped by similar nostalgic construc-tions of *The Professionals* by Comic Strip Productions, who wrote two spoofs of the series (and whose member, Peter Richardson, directed the ad, see Baird, 1998, p. 141), one in the mid-1980s, called *The Bullshitters*, and with a follow-up produced in the early 1990s. The latter focused on the death of a TV detective, and three different sets of TV detectives fighting over which era's style (from the flamboyant, crushed velvet-wearing late 1960s/early 1970s, through all-action mid-1970s [*The Professionals/Bullshitters*], to laconic, austere mid-1980s [*Spender*]) in which to go about investigating the murder. Perhaps importantly in terms of making the discursive link, the actor in the advert is also in the second of these two shows, although playing a different character.

6   As Volti (1996, p. 663) points out, there is surprisingly little academic mater-ial on the rise of the car, given its importance in twentieth-century society throughout much of the world. Nevertheless, this account draws in particu-lar on a number of general critical works which are worth mentioning at this stage. Of particular importance are Sachs (1992a), Wolf (1996), Ross (1995), Gorz, 'The Social Ideology of the Motor Car', in (1980), Flink (1975, 1988), Rupert (1995), McShane (1994), Schwartz Cowan (1997), Freund and Martin (1993, 1996).

7   Wernick (1991, pp. 67–70) suggests that from about 1980 onwards, the car was displaced as the symbol of progress by the computer. Speed/acceleration is still the predominant theme, however, which links both and enables the shift from one to the other. For a fuller treatment of the relationship between cars and growth, see Paterson (2000a).

8   This episode is mentioned widely. The fullest treatments are in Gordon (1991), Dunn (1981), and St Clair (1986). For others who discuss it, see Wolf (1996, p. 84), Hamer (1987, p. 22), Wajcman (1991, p. 28), Freund and Martin (1993, p. 135), UNCTC (1992, pp. 57–9), McShane (1994, p. 115). For a fuller treatment of these four themes, see Paterson (2000a).

9   Elton (1991) also focuses on the inequalities produced by car-centred trans-port, with one of his two main characters having cerebral palsy, and there-fore unable to drive and disabled by the car-centred system, and the other having lost her legs when hit by a car, but then reliant on a car because of

the inadequacies of London's public transport for those in wheelchairs. His other theme is corrupt relations between government ministers and the roads lobby.

10   The exception of course is the question of the suffrage, where the benefits can be claimed for a much broader group of women.

11   Although it is always dubious to conflate women and children, children's independent mobility has also been reduced by the car, as parents increasingly prevent children from going across towns and cities on their own because of fears about their safety on the roads (Hillman *et al.*, 1990, cited in Tiles and Oberdiek, 1995, p. 135).

12   Although this is always problematic, I shall stick to a narrow definition of 'environmental' here. I have discussed above some of the aspects of inequality produced by a car-dominated system, for example around class and gender. The reorganisation of urban space that a car-based system effects is of course a crucial form of environmental change in terms of people's direct experience of the space around them. The substantial number of deaths and serious injuries caused by cars, should also in many ways be considered an environmental problem. In the UK, for example, more people have been killed on the roads since 1945 than were killed in the Second World War on active service (Hamer, 1987, p. 2; for general figures on this see Wolf, 1996, pp. 201–4). Here, however, I deal only with aspects of the 'environment' involving pollution and resource consumption.

13   'Claimed to be', since catalytic converters only work fully when engines are warmed up, i.e. after four-to-five miles, but the significant majority of car journeys are under two miles. Thus their effect on actual emissions is much smaller. They have, however, led to a decrease in NOx and CO emissions, although as the IEA points out, this will be offset in the early twenty-first century by overall traffic growth, particularly of trucks (IEA, 1993, p. 31). On the other hand, they also have their own problems. They decrease fuel efficiency by a small amount and therefore marginally increase CO_2 emissions, and greatly increase nitrous oxide emissions (Wolf, 1996, pp. 174–7). They also produce small quantities of platinum and palladium, themselves with known connection to health problems (ibid.).

14   Toad is also cited by McShane (1994, p. 144) as an important example of cultural hostility to the car in the early twentieth century.

15   On opposition in the UK in this period, see Hamer (1987, chs 5 and 6). The opposition centred on plans to build inner-city motorways in London, and the expansion of the motorway programme in the early 1970s. Although in smaller numbers than in the US, the opposition was also expressed in books, see, for example, Aird (1972).

16   The large volume of US literature on this from the mid-1960s to the mid-1970s is a testament to this. For a review, see Davies (1975) or Flink (1972, ch. 7; 1988, ch. 20). The earliest was perhaps Lewis Mumford's *The Highway and the City* (1963). For a selection of others, see Leavitt (1970), Owen (1972), Mowbray (1969), Schneider (1971), Buel (1972) or Flink (1975).

17   For example Smith (1997, p. 350, note 32) citing *The Guardian*, 1 March 1996.

18   Seel (1997). For other analyses of roads protests in Britain, see for example McKay (1996), Smith (1997), Doherty (1997, 1999), Welsh and McLeish

(1996). The analyses by Seel and McKay in particular support the interpretation given here.

6   Fast Food, Consumer Culture and Ecology

1   For a general overview of the trial, see Vidal's book on it (1997), or the McLibel Support Campaign's website, at http://www.McSpotlight.org. For shorter accounts, see Rowell (1996, pp. 353–5) or Beder (1997, pp. 68–9).

2   Throughout the trial, many of the claims made by McDonald's witnesses and lawyers centred on the question of whether they broke the law. For its representatives, working within the law, were above criticism, or so they implied. For the defendants, this was irrelevant; McDonald's could be working legally yet unethically. The law is designed to protect their interests anyway, and so is not a good guide to ethical action. For analyses which focus on McDonald's employment practices, see Transnationals Information Centre (1987) or Gabriel (1994).

3   Ritzer (1996, p. 37). See also Love's history of the McDonald's company (1987). For a similar account of rationalisation of Burger King, see Reiter (1996, especially ch. 6) and more generally on fast-food operations, see Ball (1992, pp. 65–82).

4   Caputo (1998) certainly provides evidence that such cultural imperialism is not absent from academic texts on the subject. Similarly to the characters in *Pulp Fiction*, he shows, perhaps inadvertently, that McDonald's is understood by many Americans as a way of taming other countries through the reproduction of a certain predictable place to eat familiar to an American traveller. He describes a journey to Canterbury in southern England in 1981, before Canterbury had a McDonald's, where on their arrival in the city he and his family became progressively more frustrated at being able to find any food to eat. When they did find some, it was unfamiliar to them, and they repeatedly had to ask for extras (bun, salad, ketchup) to make their English burger and chips approximate something they recognised from McDonald's. How conscious he was of this when he wrote the story is not clear, but the account is heavily imbued with a narrative of the superiority of American culture, food, service, etc. This is, perhaps not coincidentally, connected to comments on the congested state of British roads. See Caputo (1998, pp. 40–2).

5   There is always a good counter-example to such claims. As I finish this chapter (June 1999), McDonald's has introduced the 'Lamb McSpicy' and 'McChicken Korma Naan', as if purpose-designed to give the proponents of the hybridisation thesis tasty morsels to chew on.

6   On ecological economics, see for example Martinez-Alier and Schluepmann (1990), or Daly (1992). Much of the argument connecting the economy to its biophysical context draws on the use of the notion of entropy developed by Georgescu-Roegen (see Guha and Martinez-Alier, 1997, ch. 9, for a discussion of his work).

7   For others, see Twigg (1983, pp. 23–4); Seager (1993, pp. 207–13); Fiddes (1991, especially pp. 144–62).

8 In addition, the intensive use of paper in packaging promotes (temperate) deforestation.

9 Smart (1994, pp. 175–8) makes similar claims, that the globalisation of food styles, producing what he terms the 'gastro-tourist', helps to conceal the costs of the industrial food system, and the unequal distribution of food across the world.

10 That chapter is instead devoted to a discussion of the driving forces (economic, cultural and social) behind McDonaldisation, and to alternative interpretations of contemporary social trends, discussing theorists of post-industrial society (Daniel Bell), post-Fordism, and postmodernism (he mentions David Harvey and Frederic Jameson at greatest length).

7 Conclusion: Globalisation, Governance and Resistance

1 The main literature to which I refer here is Lipschutz with Mayer (1996), the title of which is the subtitle to this section, Lipschutz (1997) and Wapner (1996, 1997). In a slightly different theoretical context, drawing on critical theoretic accounts of political community, especially by Linklater (1990; 1992), Samhat (1997) develops similar arguments, using climate change as one of his case-studies. For an account closer to that developed here, focusing on debates about TNCs and globalisation, see Newell (forthcoming 1999).

2 Lipschutz is, however, keen to emphasise that the functionalism here is not that of the functionalists of the 1960s (Mitrany, Haas). The functionally specific nature of emerging patterns of governance is not conceptualised as an emerging pattern of political integration as in earlier models – it does not lead to a transcendence of existing structures of government. States still exist in Lipschutz's model. Rather, functionally specific networks of governance for him are the consequence of innovation – the response of actors to the inability of traditional models of government to respond to specific problems, and the emergence of networks to learn how to respond more effectively (Lipschutz, 1997, pp. 87–8).

3 Lipschutz erroneously suggests that 'the political implications of such a process have not been given much serious thought' (1997, p. 86). Such an impression may be correct in North America; it is far from true in Europe, where the literature debating the implications of globalisation is enormous, and growing daily. Much of the literature in this debate is concerned either to show that such measures of integration do not add up to something warranting the term 'globalisation', and/or that globalisation does not mean that there is no alternative to neoliberalism. For a selection of such literature, see Hirst and Thompson (1996), Weiss (1997), Zysman (1996), Jones (1995), Ruigrok and van Tulder (1995). For (in my view) more nuanced accounts, see Palan and Abbott with Deans (1996), Kofman and Youngs (1996), or Held *et al.* (1999).

4 This is not to say that globalisation cannot be measured in part using such accounts of integration. But such measures will inevitably be heavily debated, turning as they do on the interpretation of statistics, the meaning of which will always be disputed. Conceptualising globalisation as capitalist

reconsolidation at global levels means, however, that accounts of globalisation need not rely on such statistical evidence.

5 This argument is given most fully in Saurin (1994). I give it more space than I do here in Paterson (1999).

6 Pasha and Blaney (1998) offer a critique of Lipschutz and others which starts from a similar premise to mine here, but ends up in a rather different direction. They argue that rather than offering a democratising potential in global politics, the actors making up global civil society as identified by Lipschutz, Wapner and others are engaged in supporting, in an oligarchic fashion, ongoing global concentrations of power. Much of their argument derives from an alternative theoretical account of civil society drawn ultimately from Marx (whereas Lipschutz's conception is liberal). For me, Pasha and Blaney underestimate the resistive potential of many of the diverse movements which could be said to make up global civil society.

7 But see, for example, Rangan (1996) for an alternative interpretation of such resistance. Rangan suggests that rather than opposing development itself, such protestors are trying to promote a more inclusive, participatory form of development.

8 Douglas (1997). See also Kuehls (1996, pp. 80–9), Escobar (1995, p. 10). All three writers draw heavily on Foucault for these accounts of power and discourse with respect to globalisation and/or development.

9 On how such resistance is simultaneously reconstructive in the context of the roads protests, see Seel (1997).

10 Helleiner suggests (1996, pp. 63–4) that it was Dubos (1981) who coined this phrase. The phrase does, however, have an alternative lineage from the mainstream environmentalism of the early 1970s. Situationist Raoul Vaneigem used it in his *The Revolution of Everyday Life* (1983), originally published in 1967 (in 1967 French edition, p. 185; see Marcus, 1990, p. 239). While, as Marcus notes, the phrase is now a common bumper sticker and 'the person who bought it will never know' the origins of the words this is nevertheless an interesting context for the emergence of Green politics. The concerns of Debord and other Situationists have many affinities with those of Greens.

Bibliography

Adams, Carol (1990) *The Sexual Politics of Meat: a Feminist-Vegetarian Critical Theory*, Cambridge: Polity.

Adams, Carol (1991) 'Ecofeminism and the Eating of Animals', *Hypatia*, 6, 1.

Adler, Emanuel (1997) 'Seizing the Middle Ground: Constructivism in World Politics', *European Journal of International Relations*, 3, 3, pp. 319–63.

Agarwal, Anil, and Sunita Narain (1989) *Towards Green Villages: a Strategy for Environmentally Sound and Participatory Rural Development*, New Delhi: Centre for Science and Environment.

Agriculture Committee (1998a) *Flood and Coastal Defence, Volume 1, Report and Proceedings*, HC707-1, London: The Stationery Office.

Agriculture Committee (1998b) *Flood and Coastal Defence, Volume 2, Memoranda of Evidence*, HC707-2, London: The Stationery Office.

Aird, Alisdair (1972) *The Automotive Nightmare*, London: Hutchinson.

Aitken and Howard (n.d.) *Marine and Construction Timbers Pack*, Mayfield, East Sussex: Aitken & Howard.

Alfino, Mark (1998) 'Postmodern Hamburgers: Taking a Postmodern Attitude toward McDonald's', in Mark Alfino, John Caputo, and Robin Wynyard (eds) (1998) *McDonaldization Revisited: Critical Essays on Consumer Culture*, Westport CN: Praeger, pp. 175–90.

Alfino, Mark, John Caputo, and Robin Wynyard (eds) (1998) *McDonaldization Revisited: Critical Essays on Consumer Culture*, Westport CN: Praeger.

Anderson, Benedict (1983) *Imagined Communities*, London: Verso.

Andresen, Steinar and Willy Ostreng (eds) (1989) *International Resource Management: the Role of Science and Politics*, London: Belhaven Press.

Appfel-Marglin, Frédérique and Stephen Marglin (eds) (1990) *Dominating Knowledge: Development, Culture and Resistance*, Oxford: Clarendon Press.

Appleyard, Brian (1994) 'Big Mac vs Small Frys', *The Independent*, 4 July, Section 2, p. 1.

Athanasiou, Tom (1996) *Divided Planet: the Ecology of Rich and Poor*, Boston: Little, Brown and Co.

Atkinson, Adrian (1991) *Principles of Political Ecology*, London: Belhaven.

Axelrod, Robert (1984) *The Evolution of Cooperation*, New York: Basic Books.

Baird, Nicola (1998) *The Estate We're In: Who's Driving Car Culture?*, London: Indigo.

Baldwin, David (1993) 'Neoliberalism, Neorealism, and World Politics', in Baldwin, D.A. (1993) *Neorealism and Neoliberalism: the Contemporary Debate*, New York: Columbia University Press, pp. 3–25.

Ball, Stephen (1992) 'Fast-Food Technology and Systems of Operation', in Stephen Ball (ed.) *Fast-Food Operations and Their Management*, Cheltenham: Stanley Thames, pp. 65–82.

Bandyopadhyay, J. and Shiva, V. (1987) 'Chipko: Rekindling India's Forest Culture', *The Ecologist*, 17, pp. 26–34.

Banks, Michael (1984) *Conflict in World Society*, Brighton: Wheatsheaf.

Banuri, T. and F. Apffel-Marglin (eds) (1993) *Who will Save the Forests? Political Resistance, Systems of Knowledge, and the Environmental Crisis*, London: Zed Books.

Barbier, Edward, Joanne Burgess, Josh Bishop, and Bruce Aylward (1994) 'Deforestation: the Role of the International Trade in Tropical Timber', in Brown, Katrina and David Pearce (eds) *The Causes of Tropical Deforestation*, London: UCL Press, pp. 271–97.

Bartelson, Jens (1995) *A Genealogy of Sovereignty*, Cambridge: Cambridge University Press.

Bartlett, Robert, Priya Kurian and Madhu Malik (eds) (1995) *International Organizations and Environmental Policy*, Westport CN: Greenwood Press.

Baudrillard, Jean (1988) *America*, London: Verso.

Beardsworth, Alan and Teresa Keil (1992) 'The Vegetarian Option: Varieties, Conversions, Motives, and Careers', *Sociological Review*, 40, 2, pp. 252–93.

Beardsworth, Alan and Teresa Keil (1997) *Sociology on the Menu*, London: Routledge.

Beck, Ulrich (1995) *Ecological Politics in an Age of Risk*, Cambridge: Polity.

Beder, Sharon (1997) *Global Spin: the Corporate Assault on Environmentalism*, Dartington: Green Books.

Belasco, Warren (1989) *Appetite for Change: How the Counter-Culture Took on the Food Industry, 1966–1988*, New York: Pantheon.

Bell, David and Gill Valentine (1997) *Consuming Geographies: We are Where We Eat*, London: Routledge, London.

Bellos, Alex (1995) 'McLibel Case Means Slow Service for Fast Food Chain', *The Guardian*, 28 June, p. 8.

Berkes, Fikrit (ed.) (1989) *Common Property Resources: Ecology and Community-Based Sustainable Development*, London: Belhaven.

Berman, Marshall (1982) *All That is Solid Melts into Air: the Experience of Modernity*, London: Verso.

Bernard, Mitchell (1997) 'Ecology, Political Economy and the Counter-Movement: Karl Polanyi and the Second Great Transformation', in Stephen Gill and James Mittelman (eds) *Innovation and Transformation in International Studies*, Cambridge: Cambridge University Press, pp. 75–89.

Bernauer, Thomas (1995) 'The Effectiveness of International Environmental Institutions: How We Might Learn More', *International Organization*, 49, 2, pp. 351–77.

Birch, Charles and John B. Cobb (1981) *The Liberation of Life: from the Cell to the Community*, Cambridge: Cambridge University Press.

Blowers, Andrew (1993) *Planning for a Sustainable Environment*, a report to the Town and Country Planning Association, London: Earthscan.

BMT Limited (1994) *Beachy Head to South Foreland Shoreline Management Plan, Consultation Draft*, Eastbourne: Eastbourne Borough Council.

Boardman, Robert (1997) 'Environmental Discourse and International Relations Theory: towards a Proto-Theory of Ecosation', *Global Society*, 11, 1, pp. 31–44.

Bookchin, Murray (1980) *Toward an Ecological Society*, Montreal: Black Rose Books.

Bookchin, Murray (1982) *The Ecology of Freedom: the Emergence and Dissolution of Hierarchy*, Palo Alto CA: Cheshire Books.

Booth, Douglas (1998) *The Environmental Consequences of Growth: Steady-State Economics as an Alternative to Ecological Decline*, London: Routledge.

Bourdillon, F.W. (1886) 'An Account of Recent Enquiries into Coast-Erosion; with Special Reference to the Neighbourhood of Eastbourne', *Proceedings of the Natural History Society of Eastbourne*, pp. 1–12.

Brampton, A.H. and J.M. Motyka (1983) 'The Effectiveness of Groynes', in ICE (1983) *Shoreline Protection: Proceedings of a Conference Organized by the Institution of Civil Engineers and Held at the University of Southampton on 14–15 September 1982*, London: Telford, pp. 151–6.

Brenton, Tony (1994) *The Greening of Machiavelli*, London: Royal Institute of International Affairs.

Buck, Susan J. (1998) *The Global Commons: an Introduction*, London: Earthscan.

Buel, Ronald A. (1972) *Dead End: the Automobile in Mass Transportation*, Baltimore: Penguin.

Bull, Hedley (1977) *The Anarchical Society: a Study of Order in World Politics*, London: Macmillan.

Bullard, Robert (1990) *Dumping in Dixie*, Boulder CO: Westview.

Bulloch, J. and A. Darwish (1993) *Water Wars: Coming Conflicts in the Middle East*, London: Victor Gollancz.

Bunyard, P. and F. Morgan-Grenville (eds) (1987) *The Green Alternative*, London: Methuen.

Burch, Kurt (1994) 'The "Properties" of the State System and Global Capitalism', in Stephen Rosow, Naeem Inayatullah and Mark Rupert (eds) *The Global Economy as Political Space*, Boulder CO: Lynne Rienner, pp. 37–60.

Burchill, Scott (ed.) (1996) *Theories of International Relations*, London: Macmillan.

Burton, John W. (1972) *World Society*, Cambridge: Cambridge University Press.

Butts, Kent Hughes (1994) 'Why the Military is Good for the Environment', in Käkönen, Jykri (ed.) (1994) *Green Security or Militarized Environment*, Aldershot: Dartmouth, pp. 83–110.

Caputo, John (1998) 'The Rhetoric of McDonaldization: a Social Semiotic Perspective', in Mark Alfino, John Caputo and Robin Wynyard (eds) (1998) *McDonaldization Revisited: Critical Essays on Consumer Culture*, Westport CN: Praeger, pp. 39–52.

Carey, Jim (1995) 'Big Mac versus the Little People', *The Guardian*, 15 April, p. 23.

Carter, Alan (1989) 'Outline of an Anarchist Theory of History', in David Goodway (ed.) (1989) *For Anarchism: History, Theory, and Practice*, London: Routledge, pp. 176–200.

Carter, Alan (1993) 'Towards a Green Political Theory', in Andrew Dobson and Paul Lucardie (eds) (1993) *The Politics of Nature: Explorations in Green Political Theory*, London: Routledge, pp. 39–62.

Cerny, Phil (1990) *The Changing Architecture of Politics*, London: Sage.

Chambers, D. (1987) 'Symbolic Equipment and the Objects of Leisure Images', *Leisure Studies*, 2, pp. 301–15.

Chambers, George F. (1910) *EastBourne Memories of the Victorian Period, 1845 to 1901*, Eastbourne: VT Sumfield.

Chatterjee, Pratap and Matthias Finger (1994) *The Earth Brokers: Power, Politics and World Development*, London: Routledge.

Choucri, Nazli (ed.) (1993) *Global Accord: Environmental Challenges and International Responses*, Cambridge MA: MIT Press.

Choucri, Nazli (1993a) 'Introduction: Theoretical, Empirical, and Policy Perspectives', in Choucri, Nazli (ed.) *Global Accord: Environmental Challenges and International Responses*, Cambridge MA: MIT Press, pp. 1–40.

Clapp, Jennifer (1994) 'Africa, NGOs and the International Toxic Waste Trade', *Journal of Environment and Development*, 3, 2.

Colchester, Marcus (1994) 'The New Sultans: Asian Loggers Move in on Guyana's Forests', *The Ecologist*, 24, 2, pp. 45–52.

Cole, H.S.D., C. Freeman, M. Jahoda, and K.L.R. Pavitt, (1973) *Thinking about the Future: a Critique of the Limits to Growth*, London: Chatto & Windus.

Coles, Joanna (1995) 'An Altered Image of This Green and Not So Pleasant Land', *The Guardian*, 24 February 1995, p. 24.

Conca, Ken (1993) 'Environmental Change and the Deep Structure of World Politics', in Ronnie Lipschutz and Ken Conca (eds) *The State and Social Power in Global Environmental Politics*, New York: Columbia University Press.

Conca, Ken (1994) 'Untying the Ecology–Sovereignty Debate', *Millennium*, 23, 3, pp. 701–11.

Conca, Ken (1995) 'Greening the United Nations: Non-Governmental Organisations and the UN System', *Third World Quarterly*, 16, 3, pp. 441–57.

Conca, Ken, Michael Alberty and Geoffrey Dabelko (eds) (1995) *Green Planet Blues: Environmental Politics from Stockholm to Rio*, Boulder CO: Westview Press.

Connell, R.W. (1987) *Gender and Power*, Cambridge: Polity.

Counsell, Simon (1995) Letter to David Tutt, leader of Eastbourne Borough Council, Friends of the Earth, London, 4 May.

Cox, Robert W. (1986) 'Social Forces, States and World Orders: Beyond International Relations Theory', in Robert O. Keohane (ed.) (1986) *Neorealism and Its Critics*, New York: Columbia University Press.

Cox, Robert W. (1987) *Production, Power and World Order*, New York: Columbia University Press.

Cox, Robert W. (1999) 'Civil Society at the Turn of the Millennium: Prospects for an Alternative World Order', *Review of International Studies*, 25, 1, pp. 3–28.

Cronon, William (1990) *Nature's Metropolis: Chicago and the Great West*, New York: W.W. Norton.

Cummings, Claire Hope (1999) 'Entertainment Foods', *The Ecologist*, 29, 1, pp. 16–19.

Cunningham, Caroline (1992) 'Sea Defences: a Professionalized Network?', in David Marsh and R.A.W. Rhodes (eds) *Policy Networks in British Government*, Oxford: Clarendon Press, pp. 100–23.

Dalby, Simon (1992) 'Security, Modernity, Ecology: the Dilemmas of Post-Cold War Security Discourse', *Alternatives*, 17, 1.

Dalby, Simon (1996) 'Reading Robert Kaplan's "Coming Anarchy"', *Ecumene*, 3, 4, pp. 472–96.

Dalby, Simon (1998b) 'Ecological Metaphors of Security: World Politics in the Biosphere', *Alternatives*, 23, 3.

Daly, Herman (1990) 'Toward Some Operational Principles of Sustainable Development', *Ecological Economics*, 2, 1, pp. 1–6.

Daly, Herman (1992) *Steady-State Economics*, 2nd edition, London: Earthscan.

D'Anieri, Paul (1995) 'International Organization, Environmental Cooperation, and Regime Theory', in Bartlett, Robert, Priya Kurian and Madhu Malik (eds)

(1995) *International Organizations and Environmental Policy*, Westport CN: Greenwood Press, pp. 153–70.

Dankelman, Irene and Joan Davidson (1988) *Women and Environment in the Third World*, London: Earthscan.

Davies, Richard O. (1975) *The Age of Asphalt: the Automobile, the Freeway, and the Condition of Metropolitan America*, Philadelphia: J.B. Lippincott & Co.

Dean, Warren (1987) *Brazil and the Struggle for Rubber: a Study in Environmental History*, Cambridge University Press, Cambridge.

Deudney, Daniel (1990) 'The Case Against Linking Environmental Degradation and National Security', *Millennium*, 19, 3, pp. 461–76.

Devetak, Richard (1995) 'Incomplete States: Theories and Practices of Statecraft', in John Macmillan and Andrew Linklater (eds) (1995) *Boundaries in Question*, London: Pinter, pp. 19–39.

Director of Environmental Services and Chief Engineer (1995) 'Coast Protection Improvement Works: Selection of Timber', Report to Eastbourne Borough Council Environment Committee, 19 June, Eastbourne.

Dobson, Andrew (1990) *Green Political Thought*, London: Unwin Hyman.

Doherty, Brian (1997) 'Tactical Innovation in the Radical Ecology Movement in Britain', paper for the European Sociological Association Conference, Essex University, 27–30 August 1997.

Doherty, Brian (1999) 'Paving the Way: the Rise of Direct Action against Road-Building and the Changing Character of British Environmentalism', *Political Studies*, 47, 2, pp. 275–91.

Doran, Peter (1993) 'The Earth Summit (UNCED): Ecology as Spectacle', *Paradigms*, 7, 1, pp. 55–65.

Doran, Peter (1995) 'Earth, Power, Knowledge: towards a Critical Global Environmental Politics', in John MacMillan and Andrew Linklater (eds) *New Directions in International Relations*, London: Pinter, pp. 193–211.

Douglas, Ian (1997) 'Globalisation and the End of the State?', *New Political Economy*, 2, 1, pp. 165–77.

Douthwaite, Richard (1992) *The Growth Illusion*, Dublin: Lilliput Press.

Dryzek, John (1987) *Rational Ecology: Environment and Political Economy*, Oxford: Basil Blackwell.

Dubos, René (1981) *Celebrations of Life*, McGraw-Hill.

Dunn, James A. (1981) *Miles to Go: European and American Transportation Policies*, Cambridge MA: MIT Press.

Eagar, Charlotte (1997) 'Have Modem, Won't Travel', *The Guardian*, 12 March 1997, p. 6.

Eastbourne Borough Council (n.d.) *Coast Protection Scheme, Information Pack*, Eastbourne.

Eastbourne Borough Council (1992) 'Minutes of the Strategic Planning and Economic Development Sub-Committee', 24 September, Eastbourne.

Eastbourne Borough Council (1993) 'Minutes of the Strategic Planning and Economic Development Sub-Committee', 4 March, Eastbourne.

Eastbourne Borough Council (1994a) 'Minutes of the Environment Committee', 17 January, Eastbourne.

Eastbourne Borough Council (1994b) 'Minutes of the Environment Committee', 20 June, Eastbourne.

Eastbourne Borough Council (1995a) 'Minutes of the Environment Committee', 19 June, Eastbourne.

Eastbourne Borough Council (1995b) 'Minutes of the Environment Committee', 6 November, Eastbourne.

Eastbourne Borough Council (1996) 'Minutes of the Environment Committee', 15 January, Eastbourne.

Eastbourne Borough Council (1996–97) *Civic Budget*, Eastbourne Borough Council Treasurer's Office, Eastbourne.

East Sussex County Planning Department (1977) *Report on the Problems of Coastal Erosion*, Lewes: East Sussex County Council.

Eckersley, Robyn (1992) *Environmentalism and Political Theory: towards an Ecocentric Approach*, London: UCL Press.

The Ecologist (1972) *Blueprint for Survival*, Harmondsworth: Penguin.

The Ecologist (1993) *Whose Common Future? Reclaiming the Commons*, London: Earthscan.

Ekins, Paul (ed.) (1986) *The Living Economy*, London: Routledge & Kegan Paul.

Ekins, Paul (1993) 'Making Development Sustainable', in Wolfgang Sachs (ed.) *Global Ecology*, London: Zed, pp. 91–103.

Elliott, Lorraine (1998) *The Global Politics of the Environment,* London: Macmillan.

Ellul, Jacques (1964) *The Technological Society*, New York: Alfred A. Knopf.

Elton, Ben (1991) *Gridlock*, London: Sphere.

Elvin, Mark and Su Ninghu (1995) 'Man against the Sea: Natural and Anthropogenic Factors in the Changing Morphology of Harngzhou Bay, circa 1000–1800', *Environment and History*, 1, 1, pp. 3–54.

Enser, A.G.S. (ed.) (1976) *A Brief History of Eastbourne*, Eastbourne: Eastbourne Local History Society.

Environment Agency (n.d.) *Policy and Practice for the Protection of Floodplains*, Bristol: Environment Agency.

Escobar, Arturo (1995) *Encountering Development: the Making and Unmaking of the Third World*, Princeton NJ: Princeton University Press.

Esteva, Gustavo (1987) 'Regenerating People's Space', in Saul Mendlovitz and R.B.J. Walker (eds) *Towards a Just World Peace: Perspectives from Social Movements*, London: Butterworths.

Esteva, Gustavo (1992) 'Development', in Wolfgang Sachs (ed.) *The Development Dictionary: a Guide to Knowledge as Power*, London: Zed.

Esteva, Gustavo and Madhu Suri Prakash (1997) 'From Global Thinking to Local Thinking', in Majid Rahnema (ed.) with Victoria Bawtree, *The Post-Development Reader*, London: Zed, originally in *Interculture*, 29, 2, 1996.

Fantasia, Rick (1995) 'Fast Food in France', *Theory and Society*, 24: 201–43.

Featherstone, Mike (1991) *Consumer Culture and Postmodernism*, London: Sage.

Fiddes, Nick (1991) *Meat: a Natural Symbol*, London: Routledge.

Finger, Matthias (1991) 'The Military, the Nation State and the Environment', *The Ecologist*, 21, 5.

Finger, Matthias and James Kilcoyne (1997) 'Why Transnational Corporations are Organizing to Save the Global Environment', *The Ecologist*, 27, 4, pp. 138–42.

Finkelstein, Joanne (1989) *Dining Out: a Sociology of Modern Manners*, Cambridge: Polity.

Flink, James (1972) 'Three Stages of American Automobile Consciousness', *American Quarterly*, 24, pp. 451–73.

Flink, James (1975) *The Car Culture*, Cambridge MA: MIT Press.

Flink, James (1988) *The Automobile Age*, MIT Press, Cambridge MA.

Fox, Warwick (1990) *Toward a Transpersonal Ecology: Developing New Foundations for Environmentalism*, Boston: Shambhala.

Freund, Peter and George Martin (1993) *The Ecology of the Automobile*, Montreal: Black Rose Books.

Freund, Peter and George Martin (1996) 'The Commodity that is Eating the World: the Automobile, the Environment, and Capitalism', *Capitalism, Nature, Socialism*, 7, 4.

Gabriel, John (1994) *Racism, Culture, Markets*, London: Routledge.

Gartman, David (1994) *Auto Opium: a Social History of American Automobile Design*, London: Routledge.

George, Susan (1977) *How the Other Half Dies*, Harmondsworth: Penguin.

George, Susan (1992) *The Debt Boomerang: How Third World Debt Harms Us All*, London: Pluto Press.

Gibson-Graham, J.K. (1996) 'Querying Globalization', in *The End of Capitalism (as We Knew It): a Feminist Critique of Political Economy*, Oxford: Blackwell.

Giddens, Anthony (1985) *The Nation State and Violence*, Cambridge: Polity.

Giddens, Anthony (1990) *The Consequences of Modernity*, Cambridge: Polity.

Giedion, Siegfried (1949) *Space, Time and Architecture: the Growth of a New Tradition*, Cambridge MA: Harvard University Press.

Gill, Stephen (ed.) (1993) *Gramsci, Historical Materialism, and International Relations*, Cambridge: Cambridge University Press.

Gill, Stephen and David Law (1988) *The Global Political Economy*, Hemel Hempstead: Harvester Wheatsheaf.

Goldblatt, David (1996) *Social Theory and the Environment*, Cambridge: Polity.

Goldsmith, Edward (1992) *The Way: an Ecological World-View*, London: Rider.

Goodin, Robert (1992) *Green Political Theory*, Cambridge: Polity.

Goodman, David and Michael Redclift (1991) *Refashioning Nature: Food, Ecology, Culture*, London: Routledge.

Gordon, Deborah (1991) *Steering a New Course: Transportation, Energy and the Environment*, Washington, DC: Union of Concerned Scientists/Island Press.

Gorelick, Steve (1997) 'Big Mac Attacks: Lessons from the Burger Wars', *The Ecologist*, 27, 5, pp. 173–5.

Gorz, Andre (1980) *Ecology as Politics*, London: Pluto.

Gorz, Andre (1994) 'Political Ecology: Expertocracy versus Self-Limitation', *New Left Review*, 202, pp. 55–67.

Grahame, Kenneth (1908) *The Wind in the Willows*, New York: Grossett & Dunlap.

Grieco, Joseph M. (1993) 'Understanding the Problem of International Cooperation: the Limits of Neoliberal Institutionalism and the Future of Realist Theory', in Baldwin, D.A. (1993) *Neorealism and Neoliberalism: the Contemporary Debate*, New York: Columbia University Press, pp. 301–38.

Guha, Ramachandra and Juan Martinez-Alier (1997) *Varieties of Environmentalism: Essays North and South*, London: Earthscan.

Haas, Peter M. (1989) 'Do Regimes Matter? Epistemic Communities and Mediterranean Pollution Control', *International Organization*, 43, 3, pp. 377–403.

Haas, Peter M. (1990a) *Saving the Mediterranean: the Politics of International Environmental Cooperation*, New York: Columbia University Press.

Haas, Peter M. (1990b) 'Obtaining International Environmental Protection through Epistemic Consensus', *Millennium*, 19, 3, pp. 347–64.

Haas, Peter M. (1992) 'Knowledge, Power and International Policy Coordination', Special issue of *International Organization*, 46, 1.

Haas, P.M., R.O. Keohane, and M.A. Levy, (1993) *Institutions for the Earth: Sources of Effective Environmental Protection*, Cambridge MA: MIT Press.

Habermas, Jurgen (1968) *Toward a Rational Society*, London: Heinemann.

Hahn, Robert W. and Kenneth R. Richards (1989) 'The Internationalisation of Environmental Regulation', *Harvard International Law Journal*, 30, 2, pp. 421–46.

Hamer, Mick (1987) *Wheels within Wheels: a Study of the Road Lobby*, London: Routledge & Kegan Paul.

Hansenclever, Andreas, Peter Mayer, and Volker Rittberger (1996) 'Interests, Power, Knowledge: the Study of International Regimes', *Mershon International Studies Review*, 40, 2, pp. 177–228.

Haraway, Donna (1991) *Simians, Cyborgs and Women: the Reinvention of Nature*, New York: Routledge.

Haraway, Donna (1991a) 'A Cyborg Manifesto: Science, Technology, and Socialist Feminism in the Late Twentieth Century', in *Simians, Cyborgs and Women: the Reinvention of Nature*, New York: Routledge.

Hardin, Garrett (1968) 'The Tragedy of the Commons', *Science*, 162, pp. 1243–8.

Hardin, Garrett (1974) 'The Ethics of a Lifeboat', *BioScience*, 24.

Harvey, David (1990) *The Condition of Postmodernity*, Oxford: Blackwell.

Hay, C. (1994) 'Environmental Security and State Legitimacy', in M. O'Connor (ed.) (1994) *Is Capitalism Sustainable? Political Economy and the Politics of Ecology*, New York: Guilford Press.

Hayward, Tim (1994) *Ecological Thought: an Introduction*, Cambridge: Polity.

Heilbroner, Robert (1974) *An Inquiry into the Human Prospect*, New York: Harper & Row.

Held, David, Anthony McGrew, David Goldblatt and Jonathan Perraton (1999) *Global Transformations: Politics, Economics and Culture*, Cambridge: Polity.

Helleiner, Eric (1994) *States and the Reemergence of Global Finance: from Bretton Woods to the 1990s*, Ithaca NY: Cornell University Press.

Helleiner, Eric (1995) 'Explaining the Globalization of Financial Markets: Bringing States Back In', *Review of International Political Economy*, 2, 2, pp. 315–42.

Helleiner, Eric (1996) 'International Political Economy and the Greens', *New Political Economy*, 1, 1, pp. 59–78.

Hempel, Lamont (1996) *Environmental Governance: the Global Challenge*, Washington, DC: Island Press.

Henderson, Hazel (1978) *Creating Alternative Futures*, New York: Perigee Books.

Henderson, Hazel (1988) *The Politics of the Solar Age*, New York: Knowledge Systems Inc.

Hildyard, Nicholas (1993) 'Foxes in Charge of the Chickens', in Wolfgang Sachs (ed.) *Global Ecology*, London: Zed.

Hillel, Daniel (1995) *Rivers of Eden: the Struggle for Water and the Quest for peace in the Middle East*, Oxford: Oxford University Press.

Hillman, Mayer, John Adams, and John Whitelegg (1990) *One False Move: a Study of Children's Independent Mobility*, London: Policy Studies Institute.

Hirst, Paul and Grahame Thompson (1996) *Globalization in Question: the International Economy and the Possibilities of Governance*, Cambridge: Polity.

Hobbelink, Henk (1991) *Biotechnology and the Future of World Agriculture*, London: Zed.

Homer-Dixon, Thomas (1991) 'On the Threshold: Environmental Changes as Causes of Acute Conflict', *International Security*, 16, 1.

Homer-Dixon, Thomas (1993) 'Physical Dimensions of Global Change', in Choucri, Nazli (ed.) *Global Accord: Environmental Challenges and International Responses*, Cambridge MA: MIT Press, pp. 43–66.

Homer-Dixon, Thomas (1994) 'Environmental Scarcities and Violent Conflict: Evidence and Cases', *International Security*, 19, 1, pp. 5–40.

Homer-Dixon, Thomas (1998) 'Environmental Scarcity and Mass Violence', in Gearóid O'Tuathail, Simon Dalby and Paul Routledge (eds) *The Geopolitics Reader*, London: Routledge, pp. 204–11.

Horkheimer, Max and Theodor Adorno (1979) *Dialectic of Enlightenment*, London: Verso.

Hovden, Eivind (1998) 'The Problem of Anthropocentrism: a Critique of Institutionalist, Marxist, and Reflective International Relations Theoretical Approaches to Environment and Development', unpublished PhD thesis, University of London.

Humphreys, David (1996) *Forest Politics: the Evolution of International Cooperation*, London: Earthscan.

Hunt, R. (n.d.) *The Natural Society: a Basis for Green Anarchism*, Oxford: EOA Books.

Hurrell, Andrew (1994) 'A Crisis of Ecological Viability – Global Environmental Change and the Nation-State', *Political Studies*, 42, Special Issue, pp. 146–65.

Hutchings, Claire (1994) 'Back to Basics', *Geographical Magazine*, 66, 3 (March), p. 20.

IEA (1993) *Cars and Climate Change*, Paris: International Energy Agency.

Illich, Ivan (1974) *Energy and Equity*, London: Calder & Boyars.

IUCN (1980) *World Conservation Strategy*, Gland: International Union for the Conservation of Nature.

Jervis, Robert (1983) 'Security regimes', in Stephen Krasner (ed.) *International Regimes*, Ithaca: Cornell University Press, pp. 173–94.

Jessop, Bob (1990) *State Theory: Putting Capitalist States in their Place*, Cambridge: Polity Press.

Jones, R.J. Barry (1995) *Globalisation and Interdependence in the International Political Economy: Rhetoric and Reality* London: Pinter.

Kahn, James and Judith McDonald (1994) 'International Debt and Deforestation', in Brown, Katrina and Pearce, David (eds) *The Causes of Tropical Deforestation*, London: UCL Press, pp. 57–67.

Käkönen, Jykri (ed.) (1994) *Green Security or Militarized Environment*, Aldershot: Dartmouth.

Kaplan, Robert (1994) 'The Coming Anarchy', *Atlantic Monthly*, February.

Karliner, Joshua (1997) *The Corporate Planet: Ecology and Politics in the Age of Globalization*, San Francisco: Sierra Club Books.

Keohane, Robert O. (1989) *International Institutions and State Power: Essays in International Relations Theory*, Boulder CO: Westview Press.

Keohane, Robert O. (1993) 'Institutional Theory and the Realist Challenge after the Cold War', in Baldwin, D.A. (1993) *Neorealism and Neoliberalism: the Contemporary Debate*, New York: Columbia University Press, pp. 269–300.

Keohane, Robert O., Peter M. Haas and Marc A. Levy (1993) 'The Effectiveness of International Environmental Institutions', in P.M. Haas, R.O. Keohane, and M.A. Levy (1993) *Institutions for the Earth: Sources of Effective Environmental Protection*, Cambridge MA: MIT Press, pp. 3–24.

Keohane, Robert O. and Joseph S. Nye (1977) *Power and Interdependence: World Politics in Transition*, Boston: Little Brown and Company.

Kellner, Douglas (1998) 'Foreword: McDonaldization and Its Discontents – Ritzer and His Critics', in Mark Alfino, John Caputo and Robin Wynyard (eds) (1998) *McDonaldization Revisited: Critical Essays on Consumer Culture*, Westport CN: Praeger, pp. vii–xiv.

Kerr, M. and N. Charles (1986) 'Servers and Providers: the Distribution of Food within the Family', *Sociological Review*, 34, 1, pp. 115–57.

King, Alexander and Bertrand Schneider (1991) *The First Global Revolution*, New York: Pantheon.

Kirchoff, Sue (1991) 'Study Urges Cut in Meat Output to Save Environment', *Boston Globe*, 14 July.

Kliot, Nurit (1994) *Water Resources and Conflict in the Middle East*, London: Routledge.

Kneen, Brewster (1995) 'The Invisible Giant: Cargill and its Transnational Strategies', *The Ecologist*, 25, 5, pp. 195–9.

Kofman, Eleanor and Gillian Youngs (1996) *Globalization: Theory and Practice*, London: Pinter.

Kothari, Ashish (1992) 'The Politics of the Biodiversity Convention', *Economic and Political Weekly*, 27, pp. 749–55.

Kramarae, C. (ed.) *Technology and Women's Voices: Keeping in Touch*, London: Routledge & Kegan Paul.

Krämer-Badoni, Thomas (1994) 'Life without the car: an Experiment and a Plan', *International Journal of Urban and Regional Research*, 18, 2, pp. 347–56.

Krasner, Stephen D. (1983) *International Regimes*, Ithaca NY: Cornell University Press.

Krasner, Stephen D. (1983a) 'Structural Causes and Regime Consequences: Regimes as Intervening Variables', in Stephen D. Krasner (ed.) (1983) *International Regimes*, Ithaca NY: Cornell University Press.

Kratochwil, F. and J.G. Ruggie (1986) 'International Organization: a State of the Art on an Art of the State', *International Organization*, 40, 4, pp. 753–75.

Kuehls, Thom (1996) *Beyond Sovereign Territory: the Space of Ecopolitics*, Minneapolis: University of Minnesota Press.

Kurian, Priya, Robert Bartlett and Madhu Malik (1995) 'International Environmental Policy: Redesigning the Agenda for Theory and Practice?', in Bartlett, Robert, Priya Kurian and Madhu Malik (eds) (1995) *International Organizations and Environmental Policy*, Westport CN: Greenwood Press, pp. 1–18.

Laferriere, Eric (1996) 'Emancipating International Relations Theory: an Ecological Perspective', *Millennium*, 25, 1, pp. 53–76.

Laferriere, Eric and Peter Stoett (1999) *Ecological Thought and International Relations Theory*, London: Routledge.

Lambert, Audrey (1971) *The Making of the Dutch Landscape: an Historical Geography of the Netherlands*, London: Seminar Press.

Lash, Scott (1990) *Sociology of Postmodernism*, London: Routledge.

Latouche, Serge (1993) *In the Wake of the Affluent Society: an Exploration of Post-Development*, London: Zed.

Lawson, C.M. (1992) 'Nutritional and Environmental Issues Confronting Fast-Food Operations', in Stephen Ball (ed.) *Fast-Food Operations and Their Management*, Cheltenham: Stanley Thames, pp. 171–87.

Leavitt, Helen (1970) *Superhighway – Superhoax*, New York: Doubleday.

Lee, Keekok (1993) 'To De-Industrialize – Is It So Irrational?', in Andrew Dobson and Paul Lucardie (eds) (1993) *The Politics of Nature: Explorations in Green Political Theory*, London: Routledge, pp. 105–17.

Leidner, R. (1993) *Fast Food, Fast Talk: the Routinization of Everyday Life*, Berkeley: University of California Press.

Levy, M., R. Keohane and P. Haas (1993) 'Improving the Effectiveness of International Environmental Institutions', in P.M. Haas, R.O. Keohane, and M.A. Levy (1993) *Institutions for the Earth: Sources of Effective Environmental Protection*, Cambridge MA: MIT Press.

Ling, Peter J. (1990) *America and the Automobile: Technology, Reform and Social Change*, Manchester: Manchester University Press.

Linklater, Andrew (1990) *Beyond Realism and Marxism: Critical Theory and International Relations*, London: Macmillan.

Linklater, Andrew (1992) 'The Question of the Next Stage in International Relations Theory: a Critical-Theoretic Point of View', *Millennium*, 21, 1, pp. 77–98.

Lintott, John (1998) 'Beyond the Economics of More: the Place of Consumption in Ecological Economics', *Ecological Economics*, 25, 3, pp. 239–48.

Lipschutz, Ronnie D. (1997) 'From Place to Planet: Local Knowledge and Global Environmental Governance', *Global Governance*, 3, 1, pp. 83–102.

Lipschutz, Ronnie D. and Judith Mayer (1996) *Global Civil Society and Global Environmental Governance: the Politics of Nature from Place to Planet*, Albany: SUNY Press.

Litfin, Karen (1993) 'Eco-regimes: Playing Tug of War with the Nation-State', in Lipschutz, Ronnie and Ken Conca (eds) (1993) *The State and Social Power in Global Environmental Politics*, New York: Columbia University Press, pp. 94–117.

Litfin, Karen (1994) *Ozone Discourses: Science and Politics in Global Environmental Cooperation*, New York: Columbia University Press.

Litfin, Karen (1997) 'Sovereignty in World Ecopolitics', *Mershon International Studies Review*, 41, pp. 167–204.

Love, John F. (1987) *McDonald's: Behind the Arches*, London: Bantam.

Lowenhaupt Tsing, Anna (1995) 'Empowering Nature, or: Some Gleanings in Bee Culture', in Sylvia Yanagisako and Carol Delaney (eds) *Naturalizing Power: Essays in Feminist Cultural Analysis*, London: Routledge, pp. 113–43.

Luke, Timothy W. (1996) 'Liberal Society and Cyborg Subjectivity: the Politics of Environments, Bodies, and Nature', *Alternatives*, 21, 1, pp. 1–30.

Lury, Celia (1996) *Consumer Culture*, Cambridge: Polity.

Macnaghten, Phil and John Urry (1998) *Contested Natures*, London: Sage.

Macpherson, C.B. (1975) 'Capitalism and the Changing Concept of Property', in Eugene Kamenka and R.S. Neale (eds) *Feudalism, Capitalism and Beyond*, London: Edward Arnold, pp. 104–25.

MAFF (1993) *Strategy for Flood and Coastal Defence in England and Wales*, London: Ministry of Agriculture, Fisheries and Food/The Welsh Office, September.

Mahur, Dennis and Robert Schneider (1994) 'Incentives for Tropical Deforestation: Some Examples from Latin America', in Brown, Katrina and Pearce, David (eds) *The Causes of Tropical Deforestation*, London: UCL Press, pp. 159–71.

Mann, Michael (1986) *The Sources of Social Power*, vol. 1, Cambridge: Cambridge University Press.

Mantle, Deborah (1999) *Critical Green Political Theory and International Relations Theory – Compatibility or Conflict*, PhD thesis, Keele University.

Marcus, Greil (1990) *Lipstick Traces: a Secret History of the Twentieth Century*, Cambridge MA: Harvard University Press.

Marcuse, Herbert (1964) *One Dimensional Man*, London: Sphere.

Marsh, Peter and Peter Collett (1986) *Driving Passion: the Psychology of the Car*, London: Jonathan Cape.

Martell, Luke (1994) *Ecology and Society*, Cambridge: Polity.

Martinez-Alier, Juan and Klaus Schluepmann (1990) *Ecological Economics: Energy, Environment and Society*, Oxford: Blackwell.

McCay, Bonnie and James Acheson (eds) (1987) *The Question of the Commons: the Culture and Ecology of Communal Resources*, Tuscon: University of Arizona Press.

McCully, Patrick (1991), 'The Case against Climate Aid', *The Ecologist*, 21, 6, pp. 244–51.

McKay, George (1996) 'Direct Action of the New Protest: Eco-Rads on the Road', in *Senseless Acts of Beauty: Cultures of Resistance since the Sixties*, London: Verso.

McManus, Phil (1996) 'Contested Terrains: Politics, Stories and Discourse of Sustainability', *Environmental Politics*, 5, 1, pp. 48–73.

McPhee, John (1989) 'Atchafalaya', in *The Control of Nature*, London: Pimlico.

McShane, Clay (1994) *Down the Asphalt Path: the Automobile and the American City*, New York: Columbia University Press.

Meadows, Donella, Dennis Meadows, Jorgen Randers and William Behrens (1972) *The Limits to Growth*, London: Pan.

Mellor, Mary (1992) *Breaking the Boundaries: Towards a Feminist Green Socialism*, London: Virago.

Mellor, Mary (1995) 'Materialist Communal Politics: Getting from There to Here', in Joni Lovenduski and Jeffrey Stanyer (eds) *Contemporary Political Studies*, Belfast, Political Studies Association.

Merchant, Carolyn (1980) *The Death of Nature: Women, Ecology and the Scientific Revolution*, San Francisco: Harper & Row.

Mies, Maria (1986) *Patriarchy and Accumulation on a World Scale*, London: Zed.

Miles, Steven (1998) 'McDonaldization and the Global Sports Store: Constructing Consumer Meanings in a Rationalized Society', in Mark Alfino, John Caputo, and Robin Wynyard (eds) (1998) *McDonaldization Revisited: Critical Essays on Consumer Culture*, Westport CN: Praeger, pp. 53–66.

Miller, Marian (1996) *The Third World in Global Environmental Politics*, Buckingham: Open University Press.

Mische, Patricia (1989) 'Ecological Security and the Need to Reconceptualise Sovereignty', *Alternatives*, 14, 4.

Mittelman, James (1998) 'Globalisation and Environmental Resistance Politics', *Third World Quarterly*, 1, 5, pp. 847–72.

Monbiot, George (1996) 'Protest Now: Nice and Easy Does It', *The Guardian*, 8 May, p. 24.

Moore Lappé, Frances (1982) *Diet for a Small Planet*, tenth anniversary edition, New York: Ballantine.

Mowbray, A.Q. (1969) *Road to Ruin*, Philadelphia: Lippincott.

Moxton, Graeme and John Wormald (1995) *Driving over a Cliff? Business Lessons from the World's Car Industry*, Reading MA: Addison-Wesley.

Mumford, Lewis (1963) *The Highway and the City*, New York: Harcourt, Brace and World.

Myers, Norman (1994) 'Tropical Deforestation: Rates and Patterns', in Brown, Katrina and Pearce, David (eds) *The Causes of Tropical Deforestation*, London: UCL Press, pp. 27–41.

Nebbia, Giorgio (1990) *La società dei rifuti*, Bari: Edipuglia.

Newell, Peter (forthcoming, 1999) 'Environmental NGOs, Transnational Corporations and the Question of Governance', in Robin Cohen and Shirin Rai (eds) *Social Movements and Social Institutions*, London: Routledge.

Ophuls, William (1977) *Ecology and the Politics of Scarcity*, San Francisco: W.H. Freeman & Co.

Ophuls, William and Stephen Boyan (1992) *Ecology and the Politics of Scarcity Revisited*, New York: W.H. Freeman.

O'Riordan, T. (1981) *Environmentalism*, 2nd edition, London: Pion.

Ostrom, Elinor (1990) *Governing the Commons: The Evolution of Institutions for Collective Action*, Cambridge: Cambridge University Press.

Oswald, Julian (1993) 'Defence and Environmental Security', in Gwyn Prins (ed.) *Threats without Enemies: Facing Environmental Insecurity*, London: Earthscan.

O'Tuathail, Gearóid, Simon Dalby and Paul Routledge (eds) (1998) *The Geopolitics Reader*, London: Routledge.

Overy, Richard (1990) 'Heralds of Modernity: Cars and Planes from Invention to Necessity', in Mikulas Teich and Roy Porter (eds) *Fin de Siècle and its Legacy*, Cambridge: Cambridge University Press, pp. 54–79.

Owen, Wilfrid (1972) *The Accessible City*, Washington, DC: Brookings Institution.

Oye, Kenneth O. (ed.) (1986) *Cooperation under Anarchy*, Princeton: Princeton University Press.

Palan, Ronen and Jason Abbott with Phil Deans (1996) *State Strategies in the Global Political Economy*, London: Pinter.

Palo, Matti (1994) 'Population and Deforestation', in Brown, Katrina and Pearce, David (eds) *The Causes of Tropical Deforestation*, London: UCL Press, pp. 42–56.

Parker, D.J. and Penning-Rowsell, E.C. (1980) *Water Planning in Britain*, London: Allen & Unwin.

Parker, D.J. and Penning-Rowsell, E.C. (1983) 'Is Shoreline Protection Worthwhile? Approaches to Economic Evaluation', in ICE (1983) *Shoreline Protection: Proceedings of a Conference organized by the Institution of Civil Engineers and Held at the University of Southampton on 14–15 September 1982*, London: Telford, pp. 39–46.

Parker, Martin (1998) 'Nostalgia and Mass Culture: McDonaldization and Cultural Elitism', in Mark Alfino, John Caputo, and Robin Wynyard (eds)

(1998) *McDonaldization Revisited: Critical Essays on Consumer Culture*, Westport CN: Praeger, pp. 1–18.

Pasha, Mustapha Kamal and David Blaney (1998) 'Elusive Paradise: the Promise and Peril of Global Civil Society', *Alternatives*, 23, 4, pp. 417–50.

Pateman, Carole (1988) *The Sexual Contract*, Cambridge: Polity.

Paterson, Matthew (1995) 'Radicalising Regimes? Ecology and the Critique of IR Theory', in John MacMillan and Andrew Linklater (eds) *New Directions in International Relations*, London: Pinter, pp. 212–27.

Paterson, Matthew (1996) *Global Warming and Global Politics*, London: Routledge.

Paterson, Matthew (1996a) 'IR Theory: Liberal Institutionalism, Neorealism and the Climate Change Convention', in Mark Imber and John Vogler (eds) *Environment and International Relations: Theories and Processes*, London: Routledge, pp. 59–77.

Paterson, Matthew (1996b) 'UNCED in the Context of Globalisation', *New Political Economy*, 1, 3, pp. 400–3.

Paterson, Matthew (1996c) 'Green Politics', in Scott Burchill (ed.) *Theories of International Relations*, London: Macmillan.

Paterson, Matthew (1999) 'Globalisation, Ecology, and Resistance', *New Political Economy*, 4, 1, pp. 129–46.

Paterson, Matthew (1999a) 'Green Political Strategy and the State', in N. Ben Fairweather, Sue Elworthy, Matt Stroh and Piers Stephens (eds) *Environmental Futures*, London: Macmillan, pp. 73–87.

Paterson, Matthew (2000a) 'Car Culture and Global Environmental Politics', *Review of International Studies*, 26, 2, pp. 253–70.

Paterson, Matthew (2000b) 'Swampy and the Tabloids', in Ben Seel, Brian Doherty and Matthew Paterson (eds) *Direct Action in British Environmentalism*, London: UCL Press.

Perl, L. (1974) *The Hamburger Book*, New York: Seabury.

Perry, N. (1995) 'Travelling Theory/Nomadic Theorizing', *Organization*, 2, pp. 35–54.

Pethick, John (1993) 'Shoreline Adjustments and Coastal Management: Physical and Biological Processes under Accelerated Sea-Level Rise', *Geographical Journal*, 159, 2, pp. 162–8.

Plumwood, Val (1993) *Feminism and the Mastery of Nature*, London: Routledge.

Porritt, Jonathon (1986) *Seeing Green*, Oxford: Blackwell.

Porter, Gareth (1998) 'Environmental Security as a National Security Issue', in Gearóid O'Tuathail, Simon Dalby and Paul Routledge (eds) *The Geopolitics Reader*, London: Routledge, pp. 215–22.

Posford Duvivier (1994) 'For…', *Eastbourne Herald*, 26 November 1994.

Posford Duvivier (n.d., a) *Posford Duvivier*, Peterborough: Posford Duvivier.

Posford Duvivier (n.d., b) *Posford Duvivier Environment*, Peterborough: Posford Duvivier.

Princen, Thomas and Matthias Finger (eds) (1994) *Environmental NGOs in World Politics*, London: Routledge.

Princen, Thomas, Matthias Finger and Jack Manno (1995) 'Non Governmental Organizations in World Environmental Politics', *International Environmental Affairs*, 7, 1, pp. 42–58.

Probyn, Elspeth (1998) 'Mc-Identities: Food and the Familial Citizen', *Theory, Culture and Society*, 15, 2, pp. 155–73.

Programme of the German Green Party (1983) London: Heretic.

Puckett, Jim (1992) 'Dumping on Our World Neighbours: the International Trade in Hazardous Wastes, and the Case for an Immediate Ban on All Hazardous Waste Exports from Industrialised to Less-Industrialised Countries' in Fridtjof Nansen Institute, *Green Globe Yearbook 1992*, Oslo: Fridtjof Nansen Institute, pp. 93–101.

Puckett, Jim (1994) 'Disposing of the Waste Trade: Closing the Recycling Loophole', *The Ecologist*, 24, 2, pp. 53–8.

Rae, John (1971) *The Road and the Car in American Life*, Cambridge MA: MIT Press.

Rangan, Haripriya (1996) 'From Chipko to Uttaranchal: Development, Environment, and Social Protest in the Garhwal Himalayas, India', in Richard Peet and Michael Watts (eds) *Liberation Ecologies: Environment, Development, Social Movements*, London: Routledge.

Redclift, Michael (1987) *Sustainable Development: Exploring the Contradictions*, London: Routledge.

Redclift, Michael (1996) *Wasted: Counting the Costs of Global Consumption*, London: Routledge.

Reeve, Andrew (1986) *Property*, London: Macmillan.

Reiter, Ester (1996) *Making Fast Food: from the Frying Pan into the Fryer*, Montreal: McGill-Queens University Press.

Reuter (1990) 'East Eats West as Bic Mac Arrives in Moscow', *The Guardian*, 1 February, p. 20.

Rinehart, Jane (1998) 'It May Be a Polar Night of Icy Darkness, but Feminists are Building a Fire', in Mark Alfino, John Caputo and Robin Wynyard (eds) (1998) *McDonaldization Revisited: Critical Essays on Consumer Culture*, Westport CN: Praeger, pp. 19–38.

Ritzer, George (1996) *The McDonaldization of Society*, Thousand Oaks CA: Pine Forge.

Rodman, John (1993) 'Restoring Nature: Natives and Exotics', in Jane Bennett and William Chaloupka (eds) *In the Nature of Things: Language, Politics and the Environment*, Minneapolis: University of Minnesota Press, pp. 139–53.

Rosenau, James N. (1990) *Turbulence in World Politics: a Theory of Change and Continuity*, Princeton NJ: Princeton University Press.

Rosenau, James (1992) 'Governance, Order, and Change in World Politics', in Rosenau, James N. And Ernst-Otto Czempiel (eds) *Governance without Government: Order and Change in World Politics*, Cambridge: Cambridge University Press.

Rosenau, James N. and Ernst-Otto Czempiel (eds) (1992) *Governance without Government: Order and Change in World Politics*, Cambridge: Cambridge University Press.

Ross, Andrew (1991) *Strange Weather: Culture, Science and Technology in the Age of Limits*, London: Verso.

Ross, Kristin (1995) *Fast Cars, Clean Bodies: Decolonization and the Reordering of French Culture*, Cambridge MA: MIT Press.

Rowell, Andrew (1996) *Green Backlash: Global Subversion of the Environmental Movement*, London: Routledge.

Rowlands, I.H. (1995) *The Politics of Global Atmospheric Change*, Manchester: Manchester University Press.

Ruggie, John Gerrard (1993) 'International Regimes, Transactions, and Change: Embedded Liberalism in the Postwar Economic Order', in Stephen Krasner (ed.) *International Regimes*, Ithaca NY: Cornell University Press.

Ruigrok, Winfried and Rob Van Tulder (1995) *The Logic of International Restructuring*, London: Routledge.

Runyan, Anne Sisson (1992) 'The "State" of Nature: a Garden Unfit for Women and Other Living Things', in V. Spike Peterson (ed.) *Gendered States*, Boulder CO: Lynne Rienner.

Rupert, Mark (1995) *Producing Hegemony: the Politics of Mass Production and American Global Power*, Cambridge: Cambridge University Press.

Russell, Ray (n.d.) 'The Controversy Surrounding the Replacement of Eastbourne's Sea Defences', short unpublished paper (in Eastbourne Friends of the Earth files).

Russell, Ray (1995) 'Press Release', 28 February, Horam, East Sussex.

Ryle, Martin (1988) *Ecology and Socialism*, London: Radius.

Sachs, Wolfgang (1992a) *For the Love of the Automobile*, Berkeley: University of California Press.

Sachs, Wolfgang (ed.) (1992b) *The Development Dictionary: a Guide to Knowledge as Power*, London: Zed.

Sachs, Wolfgang (1993a) 'Global Ecology and the Shadow of "Development"', in Wolfgang Sachs (ed.) *Global Ecology*, London: Zed.

Sachs, Wolfgang (ed.) (1993b) *Global Ecology*, London: Zed.

Sachs, Wolfgang (1997) 'The Need for the Home Perspective', in Majid Rahnema (ed.) with Victoria Bawtree, *The Post-Development Reader*, London: Zed, originally in *Interculture*, 29, 1, 1996.

Sale, Kirkpatrick (1980) *Human Scale*, San Francisco: W.H. Freeman.

Samhat, Nayef (1997) 'International Regimes as Political Community', *Millennium*, 26, 2, pp. 349–78.

Sand, Peter (1990) *Lessons Learned in Global Environmental Governance*, Washington, DC: World Resources Institute.

Sandler, Todd (1997) *Global Challenges: an Approach to Environmental, Political, and Economic Problems*, Cambridge: Cambridge University Press.

Saurin, Julian (1994) 'Global Environmental Degradation, Modernity and Environmental Knowledge', in Caroline Thomas (ed.) *Rio: Unravelling the Consequences*, London: Frank Cass.

Saurin, Julian (1996) 'International Relations, Social Ecology and the Globalisation of Environmental Change', in Mark Imber and John Vogler (eds) *Environment and International Relations*, London: Routledge, pp. 77–98.

Schama, Simon (1987) *The Embarrassment of Riches*, London: Collins.

Scharff, Virginia (1991) *Taking the Wheel: Women and the Coming of the Motor Age*, New York: Free Press.

Schivelsbusch, W. (1986) *The Railway Journey: Trains and Travel in the Nineteenth Century*, Oxford: Blackwell.

Schmidheiny, Stephan (ed.) (1992) *Changing Course*, Cambridge MA: MIT Press.

Schneider, Kenneth (1971) *Autokind vs. Mankind*, New York: Norton.

Schumacher, E.F. (1976) *Small is Beautiful*, London: Sphere.

Schwarz Cowan, Ruth (1997) 'Automobiles and Automobility', in *A Social History of American Technology*, New York: Oxford University Press, pp. 224–48.

Seager, Joni (1993) *Earth Follies: Feminism, Politics and the Environment*, London: Earthscan.

Seel, Ben (1997) 'Strategies of Resistance at the Pollok Free State Road Protest Camp', *Environmental Politics*, 6, 4, pp. 102–33;

Shafik, Nemat (1994) 'Macroeconomic Causes of Deforestation: Barking up the Wrong Tree?', in Brown, Katrina and Pearce, David (eds) *The Causes of Tropical Deforestation*, London: UCL Press, pp. 86–95.

Shaheen, Murad (1997) 'The Influence of the Water Dispute on the Arab-Israeli Conflict', PhD thesis, Keele University.

Shiva, Vandana (1988) *Staying Alive: Women, Ecology and Development*, London: Zed.

Shiva, Vandana (1993) *Monocultures of the Mind: Perspectives on Biodiversity and Biotechnology*, London: Zed.

Shiva, Vandana (1993a) 'The Greening of the Global Reach', in Wolfgang Sachs (ed.) *Global Ecology*, London: Zed.

Shiva, Vandana and Radha Holla-Bhar (1993), 'Intellectual Piracy and the Neem Tree', *The Ecologist*, 23, 6, pp. 223–7.

Singer, Peter (1976) *Animal Liberation*, London: Jonathan Cape.

Slater, Don (1993) 'Going Shopping: Markets, Crowds and Consumption', in Chris Jenks (ed.) *Cultural Reproduction*, London: Routledge.

Smart, Barry (1994) 'Digesting the Modern Diet: Gastro-Porn, Fast Food and Panic Eating', in K. Tester (ed.) *The Flâneur*, London: Routledge.

Smith, Mick (1997) 'Against the Enclosure of the Ethical Commons: Radical Environmentalism as an "Ethics of Place"', *Environmental Ethics*, 19, 4.

Smith, Richard J. (1996) 'Sustainability and the Rationalisation of the Environment', *Environmental Politics*, 5, 1, pp. 25–47.

Smith, Steve (1993) 'Environment on the Periphery of International Relations: an Explanation', *Environmental Politics*, 2, 4, pp. 28–45.

Snidal, Duncan (1991) 'Relative Gains and the Pattern of International Cooperation', *American Political Science Review*, 85, 3, pp. 701–26.

Sontheimer, Sally (ed.) (1991) *Women and the Environment: a Reader*, London: Earthscan.

Soroos, Marvin (1997) *The Endangered Atmosphere*, Columbia: University of South Carolina Press.

Spencer, Colin (1994) *The Heretic's Feast: a History of Vegetarianism*, London: Fourth Estate.

Spretnak, Charlene and Fritjof Capra (1985) *Green Politics: the Global Promise*, London: Paladin.

St Clair, David J. (1988) *The Motorization of American Cities*, New York: Praeger.

Star, Susan Leigh (1991) Power, Technology and the Phenomenology of Conventions: on Being Allergic to Onions', in J. Law (ed.) *The Sociology of Monsters: Essays on Power, Technology and Domination*, London: Routledge.

Stewart, Cara (1997) 'Old Wine in Recycled Bottles: the Limitations of Green International Relations Theory', paper presented to the BISA Annual Conference, Leeds, December.

Stilgoe, John R. (1994) *Alongshore*, New Haven: Yale University Press.

Stokke, Olav Schram (1997) 'Regimes as Governance Systems', in Young, Oran R. (ed.) (1997) *Global Governance: Drawing Insights from the Environmental Experience*, Ithaca NY: Cornell University Press, pp. 27–63.

Szasz, Andrew (1994) *Ecopopulism: Toxic Waste and the Movement for Environmental Justice*, Minneapolis: Minnesota University Press.

Tansey, Geoff and Tony Worsley (1995) *The Food System: a Guide*, London: Earthscan.

Taylor, Bron (ed.) (1995) *Ecological Resistance Movements: the Global Emergence of Radical and Popular Environmentalism*, Albany, State University of New York Press.

Taylor, Bron (1995a) 'Popular Ecological Resistance and Radical Environmentalism', in Bron Taylor (ed.) *Ecological Resistance Movements: the Global Emergence of Radical and Popular Environmentalism*, Albany, State University of New York Press, pp. 334–54.

Taylor, Michael (1987) *The Possibility of Cooperation*, Cambridge: Cambridge University Press.

Taylor, Stephen, Sheena Smith, and Phil Lyon (1998) 'McDonaldization and Consumer Choice in the Future: an Illusion or the Next Marketing Revolution?', in Mark Alfino, John Caputo and Robin Wynyard (eds) (1998) *McDonaldization Revisited: Critical Essays on Consumer Culture*, Westport CN: Praeger, pp. 105–20.

Third World Resurgence (various authors) (1996) 'Farmer's Rights and the Battle for Agrobiodiversity', *Third World Resurgence*, No. 72/73.

Thomas, Caroline (1992) *The Environment in International Relations*, London: Royal Institute of International Affairs.

Thomas, Caroline (1997) 'Globalization and the South', in Caroline Thomas and Peter Wilkin (eds) *Globalization and the South*, London: Macmillan.

Thomas, Caroline and Darryl Howlett (1993) *Resource Politics: Freshwater and Regional Relations*, Buckingham: Open University Press.

Thrift, Nigel (1996) 'Inhuman Geographies: Landscapes of Speed, Light and Power', in *Spatial Formations*, London: Sage, pp. 256–309.

Tiles, Mary and Hans Oberdiek (1995) *Living in a Technological Culture: Human Tools and Human Values*, London: Routledge.

Tilly, Charles (1985) 'State Making and War Making as Organized Crime', in P.B. Evans, D. Rueschemeyer and T. Skocpol (eds) *Bringing the State Back In*, Cambridge: Cambridge University Press.

Tilly, Charles (1990) *Coercion, Capital and European States, AD 990–1992*, Oxford: Basil Blackwell.

Trafford, B.D. and R.J.E. Braybrooks (1983) 'The Background to Shoreline Protection in Great Britain', in ICE (1983) *Shoreline Protection: Proceedings of a Conference Organized by the Institution of Civil Engineers and Held at the University of Southampton on 14–15 September 1982*, London: Telford, pp. 1–8.

Trainer, F.E. (1985) *Abandon Affluence!*, London: Zed.

Trainer, Ted (1989) *Developed to Death*, London: GreenPrint.

Transnationals Information Centre (1987) *Working for Big Mac*, London: Transnationals Information Centre.

Turner, R.K., I. Bateman and J.S. Brooke (1992) 'Valuing the Benefits of Coastal Defence: a Case Study of the Aldeburgh Sea-Defence Scheme', in Annabel Coker and Cathy Richards (eds) (1992) *Valuing the Environment: Economic Approaches to Environmental Evaluation*, London: Belhaven, pp. 77–100.

Twigg, Julia (1983) 'Vegetarianism and the Meanings of Meat', in Anne Murcott (ed.) *The Sociology of Food and Eating*, Aldershot: Gower, pp. 18–29.

Tyhurst, M.F. (1972) 'Eastbourne's Sea Defences', unpublished paper for Society of Civil Engineers and Technicians competition, Eastbourne.

UNCTC (1992) *Climate Change and Transnational Corporations: Analysis and Trends*, New York: United Nations Center on Transnational Corporations.

United Nations (1992) *Framework Convention on Climate Change*, New York: United Nations.

Urry, John (1990) *The Tourist Gaze*, London: Sage.

Vaneigem, Raoul (1983) *The Revolution of Everyday Life*, Seattle: Left Bank, translated by David Nicholson-Smith, originally published as Vaneigem, Raoul (1967) *Traité de savoir-vivre à l'usage des jeunes generations*, Paris: Gallimard.

Victor, David, Kal Raustiala and Eugene Skolnikoff (eds) (1998) *The Implementation and Effectiveness of International Environmental Commitments: Theory and Practice*, Cambridge MA: MIT Press.

Vidal, John (1995a) 'Parallel Worlds Fight for the Globe', *The Guardian*, 11 November, p. 25.

Vidal, John (1995b) 'EcoSoundings', *The Guardian*, 6 December, p. 25.

Vidal, John (1997) *McLibel: Burger Culture on Trial*, London: Macmillan.

Viles, Heather and Tom Spencer (1995) *Coastal Problems: Geomorphology, Ecology and Society at the Coast*, London: Edward Arnold.

Visvanatham, Shiv (1991) 'Mrs Brundtland's Disenchanted Cosmos', *Alternatives*, 16, 1.

Vogler, John. (1992) 'Regimes and the Global Commons: Space, Atmosphere and Oceans' in McGrew, Anthony G. and Lewis, Paul (eds), *Global Politics: Globalisation and the Nation-State*, Cambridge: Polity.

Vogler, John (1995) *The Global Commons: a Regime Analysis*, London: Wiley.

Volti, Rudi (1996) 'A Century of Automobility', *Technology and Culture*, 37, 4, pp. 663–85.

Wackernagel, Mathias and William Rees (1996) *Our Ecological Footprint: Reducing Human Impact on the Earth*, Gabriola Island, BC: New Society Publishers.

Wajcman, Judy (1991) *Feminism Confronts Technology*, Cambridge: Polity.

Walker, K. (1989) 'The State in Environmental Management: the Ecological Dimension', *Political Studies*, 37, 1.

Walker, R.B.J. (1990) 'Sovereignty, Identity, Community: Reflections on the Horizons of Contemporary Political Practice', in R.B.J. Walker and S. Mendlowvitz (eds) *Contending Sovereignties*, Boulder: Lynne Rienner, pp. 159–85.

Walker, R.B.J. (1993) *Inside/Outside: International Relations as Political Theory*, Cambridge: Cambridge University Press.

Wall, Derek (1994) 'Towards a Green Political Theory – In Defence of the Commons?', in Patrick Dunleavy and Jeff Stanyer (eds) (1994) *Contemporary Political Studies: Proceedings of the Annual Conference*, Belfast: Political Studies Association (UK), pp. 13–28.

Waltz, Kenneth (1979) *Theory of International Politics*, Reading MA: Addison-Wesley.

Wapner, Paul (1996) *Environmental Activism and World Civic Politics*, Albany: State University of New York Press.

Wapner, Paul (1997) 'Governance in Global Civil Society', in Young, Oran R. (ed.) (1997) *Global Governance: Drawing Insights from the Environmental Experience*, Ithaca NY: Cornell University Press, pp. 65–84.

Warde, Alan (1997) *Consumption, Food and Taste*, London: Sage.

Waring, Marilyn (1988) *If Women Counted: a New Feminist Economics*, London: Macmillan.

Waters, Malcolm (1995) *Globalization*, London: Routledge.

WCED (1987) *Our Common Future: Report of the World Commission on Environment and Development*, Oxford: Oxford University Press.

Weber, Cynthia (1995) *Simulating Sovereignty: Intervention, the State, and Symbolic Exchange*, Cambridge: Cambridge University Press.

Weber, Thomas (1988) *Hugging the Trees: the Story of the Chipko Movement*, Harmondsworth: Penguin.

Weiss, Linda (1997) 'Globalization and the Myth of the Powerless State', *New Left Review*, 225 pp. 3–27.

Welford, Richard and Richard Starkey (eds) (1996) *The Earthscan Reader in Business and the Environment*, London: Earthscan.

Welsh, Ian and Phil McLeish (1996) 'The European Road to Nowhere', *Anarchist Studies*, 4, 1, pp. 27–44.

Wendt, Alexander (1987) 'The Agent-Structure Problem in International Relations Theory', *International Organization*, 41, 335–70.

Wendt, Alexander (1992) 'Anarchy is What States Make of It: the Social Construction of Power Politics', *International Organization*, 46, 2, pp. 391–425.

Wernick, Andrew (1991) *Promotional Culture: Advertising, Ideology and Symbolic Expression*, London: Sage.

Westing, Arthur (ed.) (1986) *Global Resources and Environmental Conflict: Environmental Factors in Strategic Policy and Action*, Oxford: Oxford University Press.

White Jr, Lynne (1967) 'The Historical Roots of our Ecologic Crisis', *Science*, 155, pp. 1203–7.

Williams, Marc (1999) 'The Political Economy of Meat: Food, Culture and Identity', in Gillian Youngs (ed.) *Political Economy, Power, and the Body: Global Perspectives*, London: Macmillan. Previously given as a paper at the British International Studies Association Conference, Durham, December 1996; page citations from this version.

Winner, Langdon (1980) 'Do Artifacts Have Politics?', *Daedalus*, 109, pp. 121–36.

Wolf, Winfried (1996) *Car Mania: a Critical History of Transport*, London: Pluto.

Wood, Dennis (1992) *The Power of Maps*, New York: Guilford Press.

Wood, Roy (1998) 'Old Wine in New Bottles: Critical Limitations of the McDonaldization Thesis – the Case of Hospitality Services', in Mark Alfino, John Caputo, and Robin Wynyard (eds) (1998) *McDonaldization Revisited: Critical Essays on Consumer Culture*, Westport CN: Praeger, pp. 85–104.

Worster, Donald (1994) *Nature's Economy: a History of Ecological Ideas*, 2nd edition, Cambridge: Cambridge University Press.

Wright, J.C. (1902) *Bygone Eastbourne*, London: Spottiswoode.

Wroe, (1996) 'Invasion of the Monster Munch', *The Observer*, 25 February, p. 24.

Wynne, Brian (1989) 'The Toxic Waste Trade: International Regulatory Issues and Options,' *Third World Quarterly*, 11, 3, pp. 120–46.

Wynyard, Robin (1998) 'The Bunless Burger', in Mark Alfino, John Caputo, and Robin Wynyard (eds) (1998) *McDonaldization Revisited: Critical Essays on Consumer Culture*, Westport CN: Praeger, pp. 159–74.

Young, Oran R. (1989) *International Cooperation: Building Regimes for Natural Resources and the Environment*, Ithaca NY: Cornell University Press.

Young, Oran R. (1994) *International Governance: Protecting the Environment in a Stateless Society*, Ithaca NY: Cornell University Press.

Young, Oran R. (ed.) (1997) *Global Governance: Drawing Insights from the Environmental Experience*, Ithaca NY: Cornell University Press.

Young, Oran R. (1997a) 'Rights, Rules and Resources in World Affairs', in Young, Oran R. (ed.) *Global Governance: Drawing Insights from the Environmental Experience*, Ithaca NY: Cornell University Press, pp. 1–26.

Young, Oran R. (1997b) 'Global Governance: Towards a Theory of Decentralized World Order', in Young, Oran R. (ed.) (1997) *Global Governance: Drawing Insights from the Environmental Experience*, Ithaca NY: Cornell University Press, pp. 272–99.

Young, Oran and Gail Osherenko (eds) (1993) *Polar Politics: Creating International Environmental Regimes*, Ithaca NY: Cornell University Press.

Zysman, John (1996) 'The Myth of a "Global" Economy: Enduring National Foundations and Emerging Regional Realities', *New Political Economy*, 1, 2, pp. 157–84.

Index

accumulation, *see* economic growth

Adams, Carol, 130–4

anthropocentrism, 32, 25–7, 40 ,60

automobiles *see* cars

Beck, Ulrich, 51, 80–1, 91

Berman, Marshall, 102–3, 105–6, 114–16, 138

Bookchin, Murray, 4, 44, 53, 62, 156, 165

capital accumulation, *see* economic growth

capitalism, 6, 10, 16, 29, 32, 40, 42, 45–50, 75–6, 91, 105, 114, 122, 125, 147–9, 151, 157, 159, 161

cars, 6–7, 9, 95–117, 135
  advertising 95–9, 102, 107–8, 114–15
  as democratisers, 106–7
  environmental impacts of, 110–11
  freedom and, 101, 104
  gender and, 107–9
  identity and, 105–6, 117
  naturalisation of, 99–100
  political economy of, 100–1
  race, class and, 109–10
  resistance to, 113–17
  speed and, 101–6

Carter, Alan, 54–5, 62, 165

climate change, 6, 12, 15, 28, 58, 60, 77, 110–11, 177

commodification, 47, 49, 58, 75–6, 100, 151
  enclosure of land and, 57
  of labour, 46

commons regimes, 57, 63–5, 154

Conca, Ken, 2, 41, 24, 162, 163, 164

constructivism, 13–14, 16

consumer culture *see* consumption

consumption, 6, 8, 26, 28, 32, 46, 55, 62, 105, 126–9 130, 132–7, 147–8, 163, 164, 166
  ecological critiques of, 128–9

state legitimacy and, 44

Critical Mass, 115

Dalby, Simon, 22, 39, 155–8, 162, 163

decentralisation, 10, 36, 63, 65, 142, 152
  ecoanarchism and, 62
  see also commons regimes, localism

deforestation, 77, 84–8
  debt and, 85–6
  neoliberal approach to, 86–8
  meat and, 119, 125, 133, 137

development, 54–8, 74–7, 151–2, 154
  see also economic growth

Dobson, Andrew, 36, 37–8, 61, 156

domination, 5, 8, 39, 40, 52–3, 63, 66–74, 142, 153, 157
  the state and, 44–5
  see also patriarchy, state-building

'domination of nature', 5, 8, 39, 44, 53, 66–74
  meat and, 131–2, 134
  nation–building and, 67–74
  science and, 50–1

Eastbourne, 66–94
  history of sea defences, 67, 71, 74–7
  replacement of sea defences controversy, 77–80, 82–3

Eckersley, Robyn, 35–7, 59–61, 156
  on urgency of environmental crisis, 59

eco-authoritarianism, 3–4, 8, 59, 61

ecocentrism, 35–7, 59–60

Ecologist, The, 1, 24, 51, 61, 63–5, 76, 154, 157, 159, 163, 164, 166

economic growth, 5, 7, 8, 26, 30–1, 36, 40–1, 43–4, 46–7, 56, 57, 66, 100–1, 112, 152, 166
  as systemic imperative, 44, 46–7, 100–1
  see also development